THE CHICKEN WHISPERER'S GUIDE TO
Zero-Waste Chicken Keeping

THE CHICKEN WHISPERER'S GUIDE TO Zero-Waste Chicken Keeping

REDUCE, REUSE, RECYCLE

Andy Schneider and Brigid McCrea, Ph.D.

QUARRY

Brimming with creative inspiration, how-to projects, and useful information to enrich your everyday life, Quarto Knows is a favorite destination for those pursuing their interests and passions. Visit our site and dig deeper with our books into your area of interest: Quarto Creates, Quarto Cooks, Quarto Homes, Quarto Lives, Quarto Drives, Quarto Explores, Quarto Gifts, or Quarto Kids.

First Published in 2019 by Quarry Books, an imprint of The Quarto Group, 100 Cummings Center, Suite 265-D, Beverly, MA 01915, USA.
T (978) 282-9590 F (978) 283-2742 QuartoKnows.com

Quarry Books titles are also available at discount for retail, wholesale, promotional, and bulk purchase. For details, contact the Special Sales Manager by email at specialsales@quarto.com or by mail at The Quarto Group, Attn: Special Sales Manager, 100 Cummings Center, Suite 265-D, Beverly, MA 01915, USA.

10 9 8 7 6 5 4 3 2 1

ISBN: 978-1-63159-734-3

Digital edition published in 2019
eISBN: 978-1-63159-735-0

Library of Congress Cataloging-in-Publication Data available.

Design & Page Layout: Mattie Wells : mattiewells.com

Photography: via Shutterstock

Printed in Canada

To my son, Caleb, and daughter, Lily:

I will always be in your heart.

I will love you forever.

I will always keep you safe.

Contents

Introduction

What does zero-waste chicken keeping mean to you? As a backyard chicken keeper, you're certainly aware of how much waste chickens can produce. Their food is scattered on the ground to rot. Their waterers are soiled with dirt, causing it to be unfit for them to drink. They poop. The chicken feed bag adds to your trash output.

You may already be practicing some zero-waste chicken keeping without realizing that what you're doing is beneficial. For example, many backyard chicken keepers collect chicken poop and add it to a compost bin to develop rich soil, which will then be incorporated into a garden to grow vegetables to feed their family. This practice saves the chicken keeper money by not having to purchase fertilizer, and the richness of the soil will likely increase the total harvest weight.

This book will go beyond this step, offering many more ideas that will help you implement realistic and beneficial waste-free practices. You'll be able to reduce many different types of waste from biomass and trash that ultimately end up in a landfill, and you can make chicken keeping easier, more efficient, and economical. Additionally, you will learn not to spend money on items where repurposed materials are perfectly adequate.

Sustainability has become a catchphrase over the last decade. Certain individuals want to limit the resources they use from the outside world and still provide their families with nutritious food from their backyard. You will find that the suggestions in this book will help you to use your resources in the most efficient and effective ways.

Furthermore, if you are trying to legalize chicken keeping in your area, presenting municipalities with a zero-waste plan may help overcome some concerns about backyard poultry. You may be able to persuade neighbors and leaders that it is a myth that chicken owners negatively affect the environment around them.

Some would probably claim that there is no such thing as zero-waste chicken keeping or that zero waste is totally unattainable. However, zero waste is a goal that we can all work toward by becoming aware of strategies we can implement and by applying these measures for the betterment of our homes, flocks, and farms. Great strides have been made in the backyard chicken-keeping realm to reduce negative effects on the environment. In fact, when you intentionally begin to practice zero- or low-waste chicken keeping, you'll see that raising chickens can benefit the environment in many ways.

In the following chapters, you'll learn many different ways—some more advanced, others easy—to begin implementing zero-waste chicken practices in and around your property.

The first step in this journey is research. That's why you're here. The more research you do on proper care and chicken-raising techniques, the more prepared you will be, the less time and money you will spend on unnecessary projects and items, and the more efficient your farm will run.

A YEAR WITH
Hens AND
No Waste

In this book, we have done our best to define for you the inputs and outputs of chicken production from a zero-waste perspective. We have systematically broken down the steps of chicken production to carefully analyze where you can route your typical efforts toward a more environmentally friendly approach. We must all share this Earth. Some of us choose to share our plot of land with chickens. So rather than heading out to a feed store and then plunking down a box of chicks into a new coop, look at the footprint you are leaving on this planet.

Overloading the soil to the point of water contamination or filling the local landfill with waste from your chickens can all be avoided. We would like to help you consider the steps and measures you can take to be more efficient in your chicken-keeping habits. Some of these measures may mean more work, other times it may mean that you need to be more creative. When it is important for you to know about sources or ideas that may lead to waste, then we will give you a little nudge in the form of a "Waste Alert" so that you can be prepared to make a decision that will require some additional thought.

This chapter offers an overview of a lot of the issues you'll need to tackle as you begin to raise your chickens. We will touch on not just the stages of chicken raising but also, bedding, water, diet, and biosecurity.

Start AT THE Beginning

When you start planning chicken keeping with a zero-waste approach in mind, you first need to think about the true beginning. That means you have no chickens on hand and you are going to get started by bringing birds onto your property. It may not necessarily be the first time you have ever raised chickens, or perhaps not even the first time that chickens have ever been on your property, but you need to start at the beginning.

How will you start your flock? Some people have access to an incubator and can find someone to supply fertilized eggs for a hatching project. The vast majority of people would rather leave the slings and arrows of hatching to someone else. Thankfully, that means that to start this venture, all you need to do is phone in an order to a chick hatchery. There are several to choose from across the country so that makes this part of the process much easier.

A reputable hatchery is one that is a member of the NPIP (National Poultry Improvement Plan) program. This is a program in which hatcheries test their equipment and the related breeder flocks for diseases that can kill chicks. Historically speaking, there used to be problems with several different diseases being passed from hens to the chicks; when chicks were delivered by mail, they would be sick.

Sick chicks would die soon after delivery, an all-around bad situation, or they would pass along the diseases to chickens already on the farm, causing them to be sick or even die. NPIP tests for those diseases and today has even added some of the newer ones such as avian influenza. Hatcheries participate on a volunteer basis. Closely examining the programs in which hatcheries participate will help you decide if they are going to provide healthy chicks for your venture. The more NPIP programs that a hatchery participates in, the more rigorous their testing of breeders, which is to your benefit

Almost everyone who thinks about keeping chickens is focused on eggs, but there are actually three distinct zero-waste stages. At each stage, we can implement different strategies. The first stage is brooding. The second stage is pullet rearing, which

> You can repurpose bottles or Mason jars as gravity-fed waterers.

refers to raising a young, sexually immature female chicken. This stage differs from the third stage, egg laying, because you do not need nest boxes since the pullets are not laying eggs. Also, the diet of the pullet is different in that they do not need the level of calcium demanded by a laying-hen diet. If you are raising chicks without a mother hen, you will be brooding for 8 weeks. You'll then move into the pullet growth stage where you are building the strong foundation upon which your hen must live for the rest of her years. Skimping here can create problems down the road. This pullet stage lasts from week 9 until about week 16 to 22. Why is there such a big difference? Well, the Single Comb White Leghorn will likely come into lay at 16 to 18 weeks of age, whereas a strain of sex-linked brown egg layer will begin to lay at 20 to 22 weeks of age. The Single Comb White Leghorn is a lighter-bodied

breed that is accustomed to hot climates. Because they are smaller bodied, they therefore reach maturity faster. That means most of the breeds from the Mediterranean region will follow a somewhat similar growth and maturity pattern. However, the Single Comb White Leghorn has historically been bred many more years for egg production and as such, they come into lay at minimum a couple of weeks before most of the larger-bodied dual-purpose breeds. The sex-linked brown egg layers have also been selected for egg laying but are closer to dual-purpose breeds in body size. The main difference between these two strains of egg producers is body size. Bigger birds such as the sex-linked brown egg layer take longer to reach a mature body size before egg production.

If your goal is fast-growing meat chickens, then we can address that easily by shortening the brooding period from 6 to 8 weeks down to just a couple because your chickens will grow so fast. Your meat chickens will then be operating under the principles of management more like that of the growth stage for pullets. The big difference will be that you will be done in 6 to 8 weeks whereas that is just the brooding period for egg-laying chicks.

If slow-growing meat chickens is your aim, then any of the strains developed for that purpose will help you reach your goals. Note that we did not mention dual-purpose chickens. That is a different topic entirely. The slow-growing meat strains will be done in about twice the time it takes to get a fast-growing chicken to market weight. A dual-purpose breed of chicken is even slower, and it may take two and one-half or even three times as long to grow.

"Dual-purpose" is a traditional term. It means that the hens do a fair job of egg laying while the males are sufficiently sized at maturity for meat production. Dual-purpose chickens are not refined or specialized either for meat or egg production and, therefore, are not superstars for either purpose. So why consider dual-purpose breeds? Well, sometimes they are prettier, and that is important for some people in their decision making. Their genetics are broader in that they are less specialized, which is what some people find important with regard to breeds that are considered heritage. The taste of a heritage-breed meat bird is slightly different than that of a production strain, but that is mostly due to the older age of the chicken at the time of processing. Some dual-purpose breeds are critically endangered according to the Livestock Conservancy, an organization focused on conserving and preserving livestock breeds, and some people find that to be an important consideration.

So if you are more a steady-as-ye-goes farmer, and in this for the long game, then slow-growing meat bird strains or dual-purpose cockerels may meet your needs. This means you will brood your chicks for about the same time period as a layer, and you will harvest your chickens during the growth phase, somewhere between 12 to 18 weeks of age. Get ready to have deep pockets, though, because preliminary research into dual-purpose birds on modern feeds indicates a seriously poor feed conversion ratio. Regardless, with slower-growing strains and breeds you will buy many bags of feed before your chickens are ready to feed you!

Chicks
(DAY 1 TO WEEK 8)

Let's start with chicks from a hatchery. You will receive chicks in the mail in a transport box. The chicks will be inside, and you will need to put them into a brooder straight away so that they stay warm, fed, and hydrated. We go into the details of zero-waste brooding in chapter 5 when we discuss creative and functional housing for your flock.

Regardless, after those chicks are warm and well cared for in your brooder, you are left with a box and a box liner. Keep these items! The material inside the transport box can be replaced if it is too far gone (covered in droppings). But for the time being, hang on to both the box and the liner. During the first couple weeks of brooding, sometimes you'll end up with a weak chick that needs extra care. This transport container has adequate ventilation and space for a chick or two as it recuperates. You will still need to keep it warm in this box so keep a heat source handy.

To keep your chickens warm, use a brooder heat lamp. It's energy expensive and your power bill will increase, but the initial purchase cost is low. Bulbs eventually burn out and need replacement, or they may be dropped and break, also needing replacement. If they fall and come into contact with the bedding, the hot bulbs are a fire hazard.

There also are infrared heat sources (Sweeter Heater or EcoGlow). The initial cost is high, but they are built to last many years. And lower electricity bills and a tough design that makes them easy to clean and store are pluses. Our personal zero-waste recommendation is for the infrared heat source because it uses less electricity and is more efficient in keeping chicks warm.

If you are not going to keep your transport box and the material that lines the floor of the inside, then think about composting it. Pull out the lining and put it in your compost pile. I do not recommend scattering it about for wild birds to use in nest-building because, although you may not have chicks that are showing signs of illness early in the brooding process, it is possible for them to have an underlying disease that may affect wild birds. The box itself may be taken apart and stored flat or even put in your compost pile if it is not wax lined.

OTHER EQUIPMENT

For equipment, you will need a chick waterer and chick feeder. Metal, plastic, and glass forms of each of these are readily available at many local feed stores. If you care for them well, then they will last you for a decade or more. That means you will need to keep them clean and store them in a safe location. You may use them for your chicks only once in the life of your flock, but they then can be repurposed for a secondary use. The chick waterer and feeder can be used in your sick or quarantine pen. Any chicken that does not feel well will still need food and water, so you can just pull these standby items right off the shelf.

Selling Chicks

You can also use this chick transport container down the road if you store it properly. Perhaps you'll hatch your own chicks in a year's time and wish to sell them. This transport container can serve your needs once again after it is dusted, wiped clean, and perhaps the lining is replaced. To be clear, we are not recommending that you sell again through the postal service, especially since your box is likely to be marked with the name of the hatchery from which you purchased your chicks. Rather, this transport container can safely carry the chicks from your farm to their new local home.

❯ These chicks just came out of the transport box (seen in the background). Transport boxes have good ventilation.

Pullet-rearing OR Growth Stage
(WEEK 9 TO WEEK 20)

Brooding is done! You will know because the chickens are fully feathered, which occurs between 6 to 8 weeks of age for dual-purpose breeds. Or you will know if the outdoor temperature keeps itself at or above 70°F (21°C) even at night. Chicks cannot regulate their own body temperature when they are young, so brooding helps them stay warm as they grow. Regardless of the type of chicken you raise, you will need to brood them for the first week as they learn their surroundings and food sources.

Now your chickens are headed out to the coop. There are many different types of feeders and designs. You will need a feeder with enough space for all your chickens to access the feed at one time. Each chicken will need about 2 inches (5 cm) of feeder space. If your chickens are restricted in their access to feed, either by inadequate feeder length or by a bullying hen, then you will be weakening them when they need growth the most. Remember, fresh, clean, cool water is as important as a nutritionally balanced diet, so plenty of feeder and waterer space needs to be available for your growing flock.

Feeders can be made of wood, plastic, and metal. Many designs do not, however, prevent feed wastage. If there is room for a significant leap in poultry equipment design over the last 60 to 80 years, then it is recovering feed that is lost to "billing." Billing is when chickens scoop feed out of the feeder onto the floor of the coop, where is sits uneaten. We hope in the future that someone designs a feeder where there is a way to counteract this waste.

> These chickens are sifting through garden waste much in the same manner as do gleaners.

SOLVING THE BILLING ISSUE

Billing is a problem that can be reduced by raising the feeder up to the height of the backs of your chickens, but for young chickens that height changes every week. So that means you will need to adjust the height of your feeders weekly. Hanging a tube feeder is an excellent solution. I recommend one to two tube feeders inside the coop, depending on the size of your flock. I also recommend metal chains and large "S" hooks so that you need buy these items only once. Tube feeders made of metal can last for decades! I have even seen some in antique stores, and I know that they would work as well today as the day they were made.

Waterers are made of plastic or metal and the antique ones are made of glass and pottery. All tend to be easy to clean as long as you can get inside them with your hand and a scrub brush. Plastic wears out faster so this is a Waste Alert. Plastic gets dropped (thus broken and unusable!), or it becomes brittle from exposure to sunlight. That is not to say that glass or pottery will not break if it's dropped. Plastic also cracks easier than glass when it's on a heated water base in the winter. This is because of the expansion of freezing water and the heating of the heater base takes its toll on the flexibility of plastic.

DEALING WITH DROPPINGS

One of the first outputs from your pullets is going to be droppings. There are two ways to tackle them. The first is a targeted approach where you pick up the droppings daily and you are done in 5 minutes. You then dispose of the droppings in a compost pile. Composting is discussed in more detail in chapter 3. The compost will contain both the droppings and the bedding materials so that makes a nice blend.

Some people prefer not to bend over in their daily manure pick-up duties. You may choose to use a doggie pooper scooper, which are metal and cleanable. When it's full, you can dump the contents into a 5-gallon (19-liter) bucket and then empty it into your compost bin. The biggest mess is under the perches, so a droppings board that can be pulled out and scraped means less work. A tin can with holes punched in the bottom can be filled with sawdust and sprinkled over the droppings board to make clean-up easier.

The second way to handle droppings is to keep them mixed in with the litter and periodically clean out the bedding plus manure. This is a little more work, but it may not be necessary for a month or more depending on the size of your coop and the number of chickens you keep. You will need to compost this material in the same manner as you would with spot cleaning.

DEALING WITH FEATHERS

Another more obscure output of the chicken's growth phase are molted feathers. Chickens go through several juvenile molts before they put on their adult feathers. The adult feathers stay on the chickens until they are ready to molt for the first time after their laying cycle ends. The task of picking up and collecting those feathers is one best left to children and can be turned into a fun morning activity. The wing and tail feathers tend to be the largest and can be used for craft projects or sold. Costumers or mask-makers, especially around Halloween or Mardi Gras, are often seeking unique or inspiring feathers for their designs.

If you are collecting the feathers by hand, then you just bend over and pick them up. Keep the ones you like and are clean enough for your crafting purposes. If the feathers are soiled, you can try to wash dirt off them and then zip them back together by hand. Regular soap and water are all you need to wash dirty feathers collected from the coop.

DEALING WITH FEED

Feed is an input and that means feed bags in some cases. Depending on the size of your flock, you may want to do bulk purchases of a grower diet, but most small flocks go with feed bags at this stage. I would normally say that the feed bags themselves are a "Waste Alert" but if you proceed to chapter 2 and look at the recycling idea, then we think you may be pleasantly surprised. This is all we are going to say about chicken diets for the time being since chapter 2 is entirely devoted to feeding chickens.

Laying Hens

(WEEK 21 TO THE END OF THEIR FIRST LAYING CYCLE)

When your chickens start to lay eggs, then you are going to need to think about a couple of additional factors. Your feed will change to the laying hen diet from the grower diet. The equipment can stay the same, however, which is a savings. You will need to invest in nest boxes, which we will cover in more detail in this chapter. You will have eggs coming out of your ears soon enough!

You will need something to carry the eggs from inside the coop to the house. We like to use a plastic-coated wire basket, which is available in several sizes and colors. We like the plastic-coated wire because it can be disinfected weekly. We avoid fabric aprons or wooden baskets because they are harder to clean and may cause cross-contamination between batches of eggs.

Once back in the house, you will need to clean and inspect your eggs. Soap and a dedicated egg sponge can do spot cleaning as needed. You should never immerse eggs in water, even if it seems appropriate to use a wire basket in this way. Pass dirty eggs through warm running water and scrub with soap as needed. Lay your eggs out to dry and then pack them into egg flats or egg cartons. Place the cartons or flats in the refrigerator.

> Fresh eggs from your chickens can come in a variety of shell colors. Here we see white, cream, and light brown.

STORING EGGS IN FLATS

Not interested in buying new paper-pulp egg cartons? Perhaps your eggs are just for your family? Then keep an old egg carton around for just this purpose. But if you have soon exhausted your supply of egg cartons, then invest in plastic egg flats. They are used in incubators and can be stacked on top of one another. You can find them at hatchery supply stores or perhaps your local feed store carries

them. They hold about thirty eggs, depending on the size and design. Because they are plastic, they are easy to clean and disinfect.

For more information about what to do with eggs if you find that your flock is exceeding your family's current egg-eating needs, see page 49. You may find out that you have more options available to you than you thought.

Bedding

It's good to start thinking about bedding early on, as you will need to consider the source material and how you will get it. Once you have this material, you will need to store it. Let us go through some of these details.

One of the first decisions you will need to make is choosing the kind of bedding material to use. The most common bedding material is softwood shavings such as pine or aspen. Sawdust is also a suitable bedding material, but young poultry, such as ducks, may mistake it for food and consume it, leading to health issues. Most feed stores have shavings or sawdust for sale but these are sold in bags. This is a Waste Alert. If you are buying shavings by the bag, then you will need to dispose of the packaging material, which may either be plastic or paper. Choose paper, which is preferred because it is compostable.

Sand is also a suitable bedding material, except for brooding as it retains cold and may chill the chicks. It also gets very heavy if you need to do a clean-out. Use sand outdoors rather than inside a coop.

MATERIALS TO AVOID

Straw and hay have been recommended for ages, but you need to be aware that there are major flaws in using these materials. First of all, straw is not absorbent—unless you chop it to less than 1 inch (2.5 cm) in size. Then it is okay to use. One problem with this is that straw will yield more ammonia because it is not absorbent. Second, straw and hay have *Aspergillus* spores in them, which can lead to fungal infection in your birds. You should *not* brood your chicks on straw because young chicks have naive immune systems and can, in some cases, be particularly prone to *Aspergillus* infections.

Some regions of the world use by-products from other industries as poultry bedding. Rice hulls and peanut hulls are perfect examples of this practice. Corncobs were once used and may be a suitable alternative if you have access to a truckload of them. You would need to grind them up in a mill in order to get them down to a size that works for your flock, between ⅛ to ¼ inch (3.2 to 6.4 mm) in size.

You may just want to get a load of wood chips.

Once you have decided on the bedding material for your coop and brooder, then you need to think about how you will store it if you are not buying it in small batches. A large plastic storage container can hold several hundred pounds and some are sold with lids. Why use lids? This is so your source of bedding does not harbor rodents before you put it to use in your coop.

Go get a truckload of bedding material and shovel it into your storage container. Put the lid on and have confidence that you will be able to have 6 months to a year's worth of bedding whenever you need it. Invest in some basic bulk plastic storage containers that can hold half a ton. You will need a container of sufficient size to allow you to keep bedding for a year, so you only need to buy bedding and haul it home just about once a year if you are frugal. This would work fine for a flock of six to twelve chickens.

> No matter your flock size or housing arrangements, you should wear dedicated footwear and clothing in your daily routine

NOT SURE IF THE BEDDING IS RIGHT FOR YOUR HENS?

Look at their feet. Paw lesions can be seen on the undersides of the feet. Brown spots (not manure, mind you) or black spots with cracks are a sign that they have ammonia burns. You may not be cleaning out your coop often enough if this problem arises. You would have already smelled the ammonia, but this is a chicken's way of calling your bluff on clean-out.

Previously we mentioned cleaning out the bedding with a doggie pooper scooper. Replace bedding that you remove in this method at least once a week. Bedding in the coop should be several inches thick; we prefer between 4 to 6 inches (10 to 15 cm). Damage to a chicken's keel bone can occur if hens

hop off perches or nest boxes (the location inside the coop where a hen goes to lay her egg) and do not have much cushion for their landing should the landing be poorly executed.

Dust is one of the last factors to consider when choosing your bedding. You do not want a material that is super dusty. Corncobs can fall in this category during grinding. It is a good idea to keep the coop well ventilated as you spread out the new bedding. Once a month, during clean-out, it is a good idea to sweep upper surfaces to free them from dust. Crossbeams or window ledges can build up dust from the regular movements of chickens as they stir up their bedding.

NEST BOXES TO CONSIDER

There are a lot of different ways to provide nest boxes. Nest boxes can be made of wood, metal, or plastic. Many of them are cute but not practical with regard to cleaning. You will need to seal wood to make it easier to clean and because red mites like to hide in the tiny spaces of wood. So you will be denying them a place to hide when the wood is sealed. Solid sides to your nest boxes make them much easier to clean and maintain. Fewer places for dust to reside means less work during clean-out. Milk crates are often depicted as a nest box, but they are notorious for failing to hold nesting material.

The moral of the story to choosing no-waste nest boxes is to make sure they are easy to clean and can be closed. Closed? Yes, indeed! You want chickens roosting on your perches at night, not in the nest boxes. This means that the droppings that they deposit at night will not end up in the nest boxes, but rather, on the droppings board under your perches. Eggs are kept cleaner when the nest boxes are clean too. Clean eggs translate to less work in the house when you bring them in. On some models of nest boxes, all you have to do to close it is to simply fold up the perch in front and the entrance is blocked.

Some people worry about not getting out to the coop early enough in the morning to open up the nest boxes. It needs to be a regular time and getting out to the coop between 6:30 and 7 a.m. is ideal.

As much as we would like to say that there is a solution for that, there is not. The only other solution is to leave nest boxes open. Any hens you find roosting in them at night will need to be moved to perches daily until they learn not to sleep in the nest boxes. Maybe there is an inventor out there who can put a timer on a nest box that will open

and close automatically. Now, wouldn't that be something to see!

Nest boxes need a good material inside them upon which the eggs sit after they are laid. The nesting material in a nest box needs to be only 1 or 2 inches (2.5 or 5 cm) thick. The same material that you use for the bedding can also be used inside your nest boxes. However, this means that you will need to replace the contents of the nest box about as often as you do the bedding. Wood chips, milled corncobs, and hay can be retired to the compost for your no-waste solution. Nest box pads are typically a plastic square pad that goes in the bottom of the nest box. They are beneficial because they are easy to clean. All you have to do is rinse and let them dry before you reuse them. They are likely to last ten years or more, unlike the bedding choices previously discussed that must be removed and replaced once soiled.

No-Waste Biosecurity

Face it! Nobody likes thinking about biosecurity. It makes for more work than many flock owners even want to think about in the first place. Where have I been in the last three days? Did I wear these shoes to the feed store? Is that a mouse dropping!? Oh, for heaven's sake!

The days of happy-go-lucky chicken keeping like our parents and grandparents enjoyed are gone. Diseases have gone global and will just as happily end your flock as they will an entire industry in the state next door. We tend to be the cause, too! Footwear is one of the most common causes of a disease outbreak in a flock. One of the most important no-waste things we can do is to have dedicated footwear that we wear *only* into the coop. Footbaths are great and reusable once they're set up. The only Waste Alert we have to contribute on this front is the packaging for any disinfectant that you choose to have for your footbath.

The most waste-intensive practice you can have is to use disposable boot covers for guests who come onto your farm. Those are plastic that must go into a landfill, so it would be best to have no guests on your farm. Although a closed-door policy is best with regard to keeping disease off your property and out of your flock, it may not be realistic. In place of disposable boot covers, you could reuse objects such as plastic grocery store bags tied over footwear to keep your flock safe. Just be sure they don't have holes and don't start to wear through.

Hot, soapy water and elbow grease do wonders for biosecurity and cleaning around the coop. Exposure to the sun and wind after cleaning is usually all it takes for many plastic surfaces to no longer retain bad bacteria. Take everything apart and scrub, scrub, *scrub* to get things clean. Then let them dry on a warm, sunny day, and you will be pleased with the result.

Personally, I like power washers. They feel more like toys, and they make cleaning faster. I dislike cleaning, so anything to make it go faster is a good thing or I may not do it as well as I should. Disinfectants are a perfect follow-up after surfaces have dried from cleaning.

> Biosecurity requires vigilance and attention to detail so your flock stays healthy.

CONSIDER YOUR CLOTHING

Dedicated clothing (coveralls or a pair of overalls, a pair of boots, and a hat) follow along nicely with dedicated footwear. You are going to do your laundry every week anyway, so tossing in your coveralls from the coop is not much of an extra step. The same goes for dedicated equipment. Keep what is yours on your farm and, ideally, do not let others borrow your poultry equipment. *Any* equipment that is borrowed from you should be thoroughly cleaned by you upon its return.

Allowing critters to harm or even kill your chickens wastes time, money, food, and creates additional work. Keeping predators, rodents, and other critters out of your coops is a critical no-waste biosecurity step. Pets that want to eat chicken poop are taking away your chance to add it to the compost pile. Rodents are a whole other matter. The most waste-intensive measure is to use bait. Snap traps or a no-kill tin cat are a great no-waste alternative to baiting mice or rats. You can also disrupt their surroundings by removing harborage, places where they feel safe—perhaps even safe enough to reproduce. If they are kicked out of their sleeping spots, then they might move on or be more likely to be caught in your traps.

Insects are harder to deal with. You will need to check chickens regularly for external parasites. There are several acceptable methods for dealing with external parasites, the first of which is to give an affected chicken a bath. It takes time, and can be frustrating, but your chickens will thank you in the long run. Mosquitos feed on chickens and can transmit diseases such as fowl pox. Insects such as flies and mosquitoes come in because the conditions are right for them to reproduce. Keep your litter dry and flies won't breed in it. Mosquitos like standing water, so if you can dump out those sources of water, you can reduce the number of mosquitos near you.

Through the Years

Your hens will not lay forever. That is a fact. After two to three years, your laying hens will likely lay at only 75 to 85 percent of their former peak production in their first season. Their laying will steadily decrease after each molt until they lay so infrequently that you really will need to determine if they are worth keeping around. At this point, some chickens may be considered pets. You may believe your pet chickens are worth the money and care due to the joy they give you and the love you have for them. Others feel this is waste and consider alternative options at that point.

Keep in mind that your best laying hens are the ones that have been genetically selected for egg production. That means the Single Comb White Leghorn as well as any sex-linked brown-shelled egg layer. If your flock contains dual-purpose or even ornamental breeds, then your rate of egg production will be even lower.

So how long can chickens live? Ten to fifteen years is not uncommon, but their best egg production comes during the first five years. So what do you do with your hens when they are no longer earning their keep on your farm? Chapter 6 will point you in the direction of turning those hens into something that still has value to your family.

You really ought to keep only one flock on your farm at one time. This is an all-in, all-out policy. This is also a good management practice that breaks the life cycle of disease-causing organisms. If you cannot be without eggs or must start a flock of meat birds, then they should be housed completely separately and as far apart from one another on the farm as you can possibly arrange. Care for the young birds and then the older flock, and never go in reverse order. Separate the equipment, too.

THE TIMING OF REPLACING OLDER HENS

Once you have completely replaced your old flock with a new one, you will want to have some downtime before putting the younger chickens in the previous flock's coop. A downtime of at least 2 weeks after a full clean-out and disinfection is advised. Then it will be time to move the younger chickens in!

If this timing needs to coincide with the start of lay for your hens, then we advise you to think of this as being a 3-week-long process. Week 1 will be processing and clean-out. Weeks 2 and 3 will be the downtime. So if your hens begin laying at about week 16, as would be the case for some strains of Leghorn, then you will need to initiate processing the older flock when the pullets are only 13 weeks old. The stress of moving hens into a new chicken house can really throw them off so you could see a delay in the start of lay by about a week or so as they adjust to their new surroundings. Anyhow, you will need to have a calendar handy so you can count back and determine when you need to free your schedule for all this activity.

Now that you understand how the hen is likely to perform as it grows from a chick into a full-grown layer, you can start to learn about how subsequent seasons are likely to affect her.

THINKING ABOUT
Food & Water Year Round

KEEPING YOUR CHICKENS HYDRATED

Let's start with the nutrient water. During the winter, a hen that stays in lay is going to need the same amount of fresh, clean water every day to make her daily egg. But if that water is frozen, then hens tend not to break through thicker ice to access their water. A chicken's beak is an instrument for grasping seeds, not for chiseling through ice, so she may go thirsty due to a frozen waterer. Put a heated water base in the coop, and this problem is solved. If you live where the temperatures infrequently drop below freezing, you may choose to have just a secondary water source ready to switch out with fresh flowing water as needed.

In the summer, chickens that are hot will drink more water. Sometimes you can tell that the hot weather is affecting your chickens because their droppings turn just a tad bit wetter. This is because they have been drinking water to keep cool and hydrated. If you can cool your chickens, then the wet droppings problem, which can translate to wet litter, will no longer be an issue. A shade cloth secured to your chicken coop may provide them just what they need to reduce the issue of wet litter and decrease how often you'll need to replace the coop bedding.

THINKING ABOUT A DIET FOR ALL TIMES OF THE YEAR

The diet itself does not alter too much if you purchase bagged feed. All the hard work is done for you as it relates to low-cost feed formulation. Grains and ingredients are only seasonally available and then must undergo storage, but this is not a worry of yours. The only way that feed can be affected by foul weather is if the store is closed due to a hurricane or blizzard!

It's also important to think about how you are going to store your chickens' feed. Chicken feed should be kept safe from the elements and animals interested in stealing it. If you purchase a 50-pound bag (22.7 kg) or smaller, your feed will fit perfectly into a 20-gallon (55.7 liter) galvanized metal container. This will reduce the problem of spoilage due to moisture entering the bag. It will also deter many animals from accessing the bag and stealing the contents. However, some animals such as raccoons may be smart enough to remove the lid. You may have to take additional steps like securing the lid

with a bungee or even locking it with a chain. Further steps may be need if you have bears in your area as they can easily damage the can with their great strength to access the contents. Food may need to be housed in a locked barn to protect against the waste of losing food.

Many people buy treats either locally or from the internet, and that is not changed by the changing of the seasons. Ah, but yes, the greens. Chickens cannot access the nutrients that are only vaguely available in pasture grasses when those grasses become dormant. What are you to do then? Fear not! There are entrepreneurs out there who have solved this problem, and you may not even be aware of it. For more details, see page 50.

So there you have it. You know what your inputs and outputs are likely to be during the course of the average year for your flock. In the next chapter, we are going to delve into the most cost-intensive portion of chicken keeping, outside of the coop itself.

2

Feeding
Chickens

We all appreciate a good meal. Whether or not that meal includes eggs or poultry meat is entirely up to you. But not many people really think hard about the food that is on their daily plate. That is to say that not a lot of people are directly connected to their food in the same ways as a farmer or backyard flock owner.

The amount of learning that would need to take place for most people to really appreciate the food on their plates would overwhelm most if it was not provided in manageable chunks. We used to teach that in the classrooms, but that's no longer true in many cases. Thankfully, readers such as yourself want to know and appreciate where their food comes from.

This certainly extends to the food that you eat. That means the soil for your fruits and vegetables. That means the food that your animals eat. Where do you plan to source your complete and balanced ration for your animals? In this chapter, you'll learn all about how to responsibly and economically feed your chickens so they will thrive.

Finding Balance

Did you know that we know more about the diet of a chicken than any other species? Yes, indeed! Why, you ask? Well, early in the days of vitamin and mineral discovery, chickens were the research models for studies that eventually teased out information that we take for granted today. We did not always know what a vitamin was or that certain plants were the key to providing it so that our bodies, and the bodies of animals, would work properly.

And that is the key, isn't it? Imbalances lead to problems and deficiencies. With regard to commercially available poultry diets as well as bagged diets, all the work has been done for you. There's no need for you to run out and get a 4-year degree in poultry nutrition at your nearest university poultry science department.

The process of designing a balanced ration is called least-cost diet formulation. Not only does it take into account the cost of every ingredient, but it also takes into account each ingredient's nutrient content. Not all ingredients are equal; they must be balanced with others to meet the needs of the animal that will consume the ration. One of the best parts of least-cost diet formulation for chickens is that it includes waste from other industries. Bakery waste, waste products of beer-making, and even tomato pomace have been experimented with as feed ingredients. The entire poultry industry likes to look carefully at what others toss out. Will that product meet the needs of a chicken in some way? Can it be successfully integrated into a diet that will meet the needs of a chicken at a particular life stage?

All this is done in the name of efficiency. The other goal is to save money! If an ingredient that was formerly ending up in the landfill will now feed animals, then we all benefit from this zero-waste approach. So even if you are not able to grow, store, and process your own feeds for your flock, at least now you understand that bagged feed is already formulated with zero waste in mind.

Life Stages

You need to feed the right feed for the right life stage. Chicks need a diet that meets the needs of their little beaks and rapid growth rate. You will feed a layer diet only about 10 days before your hens lay their first eggs, and then you will continue feeding this diet for the remainder of their lives. You also need to feed a ration that is appropriate for the species of poultry with which you are working. For example, chickens and turkeys are different, and their bodies have different protein requirements.

A chick's diet should be at about 21 percent crude protein; higher than that and a chick's kidneys can be harmed. You will need to feed a starter diet for laying hen chicks for about 8 weeks, whereas for broiler chicks, you will feed a starter diet for only 2 weeks. Why is there such a big difference? They are two completely different birds with regards to growth genetics. One has been selected for rapid growth rate for the last 60 years and now can reach market weight in 6 to 8 weeks.

There is a growth stage between being a chick and a laying hen. This is where it gets tricky for many backyard flock owners. The ideal percentage of crude protein is 18, but few people look carefully at their feed tag labels. Some people buy what is called a "start and grow mix." This tends to have a different percentage of crude protein from what is specifically needed either for the growth or chick stage. What is being given up by choosing this type of bagged feed? Well, for backyard flock owners,

there is not a lot of published information out there. There have been no studies about how this type of diet affects growth rate, onset of lay, rate of lay, or even egg size for the ornamental or dual-purpose

chicken breeds. If the information has been collected, then it has yet to be made available to small flock owners so that they can use it for better decision making.

So for your laying hens, the best protein percentage is 16 percent. You should start feeding a laying hen diet about 2 weeks before the onset of lay. When is that going to be? Glad you asked! If you have Single Comb White Leghorns, then they will begin to lay at 16 to 18 weeks of age, so you will need to feed your laying hen diet at 14 to 16 weeks of age.

For sex-linked brown-shelled egg layers, such as production reds or production blacks, they tend to start laying 2 weeks later than the Single Comb White Leghorn. That means that they will begin laying at 18 to 20 weeks of age, and you should begin feeding a laying hen diet at 16 to 18 weeks of age. Why such a difference? Well, sex-linked brown-shelled egg layers are larger

birds and need additional time to reach physical and sexual maturity.

What about ornamental breeds such as Silkies, Ancona, or Sussex? Well, folks, the data on when they start to lay just doesn't exist. These breeds may be popular on the backyard and small flock poultry scene, but their numbers do not warrant funding the research at this time. The time that it would take to find out would be years, if not decades. Besides, the genetics of each breed have a big role to play and there are a lot of different genetics out there. If the commercial poultry industry should pick up one of these breeds for their niche market, however, then you had better believe that the work will be done to answer the questions of lay and dietary needs. For now, it is usually safe for a poultry expert to recommend that they will follow the same onset of lay as that of a sex-linked brown-shelled egg layer.

TOOLS FOR PLANNING
Low-Waste Feeding

There are computer programs that will help you design a balanced ration for your chickens. However, they are pricey and take a good measure of knowledge about the ingredients to make them work well. That said, they are recommended because they not only take into account the different aspects of the ingredients, such as amino acids or mineral content, but costs as well. Where they fall down, however, is with ingredients about which we have no data. We don't have the data because nobody has done the research on it yet. Using an untested ingredient means that your results may or may not follow performance expectations. Some new ingredients look good on paper but when they're incorporated into feeds, the overall effect may be to suppress growth or egg production rather than improve it. New ingredients are found all of the time, but the real test of whether or not research is done on a particular ingredient is in the ability of a particular researcher to write a grant that will fund the necessary research.

When it comes to the least costly diet-formulation software, consulting with a poultry nutritionist is perhaps your best bet because there are free and for-sale software programs. There are dozens of programs out there, and the one you choose will depend on your needs and ability to use a computer.

Bells and whistles come with more detailed software and least-cost feed formulation programs. If you are working with more than one formula and ingredient database, then you will most likely need to pay for software that will support your needs. Ingredient databases are where much of the value is in the case of needing more than one diet formulated. Are you beginning to see why some students spend years at college just working on poultry nutrition?

Some companies will sell you an annual license to access their proprietary software. Some software gets rather pricey, and that is a result of the time it takes to make the software user-friendly so that the product being developed is of greater value to the customer. You will need to live and breathe the software so make sure it is something you demo before buying. Additionally, least-cost formulation software is based on information provided by the National Research Council (NRC), which has compiled data on different life stages of poultry and their

developmental needs. The NRC has not updated their information on poultry in several decades but is likely going to update their data soon, which will change the formulations that we use.

What if you have a truckload of leftover watermelons that you gleaned from a nearby field? How can you determine the effect that such an ingredient would have on your feed? The water content would also be an issue for mold growth or potential fermentation. How would you store them? These are all the things that a feed mill manager and poultry nutritionist think about when exploring regional or seasonal ingredients. If you want the answers to questions like these, you'll either have to do the research yourself or pay experts to perform it for you.

FOOD SCRAPS

Here is another scenario: You obtain all of the waste products on a daily basis from a local school cafeteria. Instead of sending it to your local landfill, you bring it home 5 days a week in the back of your truck by the trash can full. Of course, you could compost it all, but would it not be better to give it to your chickens?

Well, hold on there, folks! You need to consider a couple of things. Chickens cannot digest dairy products. They do not make the enzyme lactase, which is needed in the gut to break down lactose. So you will need to make sure that the waste does not contain dairy products.

Also, the waste should not contain indigestible trash, such as plastics or wax-coated paper. So you would need to work with the kitchen so that they separate the food waste from the trash. Also, you would need to work with them to potentially include the kitchen prep waste and not just the uneaten food from breakfast or lunch.

If you are still willing to give this a go, then onward and upward! You will face looser stools from your chickens. The lessening of the dry matter in your flock's intestines will cause an increase in ammonia production, so that will need to be monitored closely. Chickens may not eat all of the waste that you provide, so you need a plan for its daily disposal when the chickens do not clean their plates.

Labor will increase and so you need to know that you can physically lift, move, and bend containers of size if you are regularly collecting food waste from any large restaurant or commercial kitchen. Containers will need to scrubbed clean daily. Bacteria, fungi, and yeasts like to grow rapidly, so you will need to clean and sanitize equipment daily. This takes time and so is rejected by most folks who also have a full-time job. Cleaning is certainly not the most fun part of everyone's day so be sure to consider the time requirement. Skimping on cleaning can lead to disease.

Most backyard chicken keepers aren't going to collect food scraps from schools or restaurants, but they probably will want to make use of their own food scraps. If you decide to go this route, serve food scraps immediately and toss any old or moldy scraps. Remember: Chickens should never depend on food scraps alone, so a balanced ration must always be available to them. And be sure to stay away from meat and dairy products.

GROWING YOUR OWN FEED

If bringing in food waste from a school or business just is not one of your ideal situations, then you may want to increase your self-sufficiency in other ways. You can grow your own poultry feed. Yes, it can be done! It is a *lot* of work, but it can be done. You will need to think a great deal about storage and yields, drying and roasting grains, rodent control programs, and equipment purchases. You will need a love for the land, the outdoors, and enough acreage to make it all happen.

So let's look at the space requirements on a per-chicken basis. The main two ingredients in a poultry diet are corn and soybeans. You will need a plot of land that is 24 by 24 feet (7 by 7 meters) to grow enough corn to feed a single laying hen. That is if you get a yield of 100 bushels (3.5 cubic meters) per acre. There are a great many factors that affect your yield. They include the strain of feed corn used, irrigation, insect and disease pressure, and weather. Get into trouble in any of these areas and your yield can go down.

As a reminder, this is not sweet corn, which is much more readily available through seed catalogs. The corn you grow to feed to your flock will be used just for this purpose: Feed corn is just for animal feed.

Once your feed corn is harvested, you will need to dry it. In the interim, you'll need to store the corn in a manner that prevents rodents from gaining access. Have we mentioned that a single mouse dropping has the potential to carry tens of thousands of salmonella bacteria? Well, it does, so let's keep the rodents at bay.

You will need to keep the corn someplace that has good airflow to let it dry out without molding. A couple of box fans running 24 hours a day for several weeks will do the trick if you leave the corn on its cob indoors in bins with wire sides or perhaps in plastic bins with slatted sides. This is assuming that you are going to store enough grain for more than just one chicken, as most small flocks consist of at least six hens.

When you harvest corn, the cob accounts for 20 percent of the weight of the bushel. We talk about what you can do with that cob in a different chapter of the book, specifically bedding (see page 25). So how do you separate the corn from the cob? Of course, you can remove the corn by hand, but there is a device called a corn sheller. It spits the cob out in one direction after removing the dried corn from the cob. The more practical type of corn sheller for a small flock owner is a hand-crank style. Be aware that shelling is dusty business, as is any step dealing with handling grain. Shelling corn is interesting work for about the first ten cobs and then it gets tedious. Perhaps this is an appropriate chore for kids or grandkids so that they can earn some spending cash?

Soybeans will need about the same amount of space. A 25- x 22-foot (7.5 x 7-meter) plot will do the trick. Soybeans, however, do not grow at exactly the same time of year as corn. The number of bushels you are likely to get from an acre of soybeans is approximately 50 (1.8 cubic meters). Of course, the same factors that affect your yield of corn are also likely to affect your soybean yield. Feel like a farmer yet?

SOYBEAN KNOW-HOW

Soybeans tend to be harvested about a 4 to 6 weeks after corn. Do you have the space to hold all that corn while you wait on the soybeans? Soybeans will need to be harvested by hand and the beans removed from the husk. They also then need to be hulled (the hull is the thin papery skin over the seed). Soybeans are great survivors. They do not like to be eaten. They contain antinutritive factors that make them indigestible. So that means you need to roast soybeans to inactivate the trypsin inhibitors.

How is that done? Well the goal is to get them to 100°C or 212°F for a certain length of time. You can boil them for 9 minutes, roast them in the oven for 2 minutes, or microwave them for 3 minutes. One of the issues with baking soybeans in the oven is that once you go over 175°F (79°C) you begin to damage isoflavones, which are one of the more desired components of soybeans. The longer the soybeans stay at certain temperatures, like the ones mentioned earlier in this paragraph, the more you modify the inherent phytochemicals. Feel like a feed-ingredient analyst yet?

MIXING AND GRINDING

Now you finally have your feed ingredients in a format that you can work with. They will need to be ground up. Don't get us wrong; whole grains are great, but they cannot be utilized by smaller chickens. A regular grain grinder will do the trick. They come in a couple of different styles, hand-crank and electric.

Remember when it was mentioned that shucking was dusty work? Well, grinding is super-duper dusty. You want to do this outside or in a garage if you are working with volumes of sufficient size to feed a flock for a week.

Once the soybeans are ground, then all you need to do is add the other minor ingredients to make a mixed feed. It will look like a mash with some larger pieces at this point; uniform size is good. With mixed feed, you will want to use a balanced ration for best results.

There are lots of recipes out there depending on what ingredients you want to use and what is available for you to use. It is a dynamic and ever-changing situation. Not all corn is as nutrient packed as the next. This is also true of other ingredients. Recipes really vary based on which region of the country you live in and what nutrients are naturally abundant in the soil and water. There are County Extension publications out there where you can research individual ingredients and see if you want to use it or can afford it. If you are in Canada, the UK, or Australia, contact your country's version of the United States Department of Agriculture for more information.

As much as people want to be fully independent and sustainable, it is still recommended that a vitamin and mineral premix for poultry be used to ensure that the chickens receive as close to a balanced ration as possible. This is doubly important for the different age groups. Get a premix that contains more calcium and phosphorous if you have hens in lay. You are not going to be able to just run out to your local feed store to buy a premix. Rather you will need to buy it in bags that range from 5 to 50 pounds (2.3 to 22.7 kg) depending on the manufacturer. If you are interested in purchasing this, there are several companies selling it. Just do an online search for "poultry diet vitamin and mineral premix."

When you are shopping for a good premix, keep in mind that it first needs to be for poultry. Second, it should be for the type of poultry you are raising, either for meat- or egg-producing birds. Third, the premix should be for the age of chickens you are working with. Last, it should be sold in volumes that you can reasonably store and at a price you can afford.

Put Your Kitchen Mixer to Work

You can use a simple kitchen mixer to achieve a uniform mix. You should add all the ingredients to the bowl and run the mixer for at least 5 to 10 minutes for each batch. You will need to add a fat of some form or else your chickens will not find the feed as palatable. You can use tallow, lard, or even poultry fat. Those are saturated fats, but you can use unsaturated fats such as corn, soy, or canola oils. The amount of fat that should be added to the mixer in an average diet is 3½ percent. Fats can go rancid if they're not handled well. Ideally, fats should be kept out of direct sunlight, in sealed containers, and used as soon as possible. Keep this in mind during the summer. Fats that go rancid reduce the effectivity of delivering the fat-soluble vitamins A, D, E, and K.

You can pretty much follow any recipe for ingredients that are available on university fact sheets or chapters in poultry nutrition books (see table at right). If you have a flock of six chickens and you mix your feed for them once a week, then you will be making 10.5 pounds (4.7 kg) of feed. You should weigh all of your ingredients before mixing.

Examples of broiler starter and caged layer peak egg production feeds

INGREDIENT	BROILER STARTER (%)	LAYER PEAK (%)
Corn, yellow	56.45	60.50
Soybean meal (47.5% CP)	27.33	21.50
Meat and bone meal (50% CP)	7.00	5.09
Meat meal (56% CP)	--	--
Bakery by-product	6.00	--
Animal/vegetable fat	1.82	3.00
Limestone (or oyster shell)	0.49	8.66
Dicalcium phosphate	0.13	0.49
Salt	0.10	0.20
Sodium bicarbonate	0.20	0.20
Copper sulfate	0.05	--
Vitamin/mineral premix	0.25	0.25
DL-methionine (99%)	0.17	0.11
L-lysine HCl (78.4% lysine)	--	--
Bacitracin-MD (50 g/lb)	0.05	--
Coban (monensin) 30 g/lb	--	--
Nicarbazin (25%)	0.05	--
Liquid mold inhibitor	0.05	--

CALCULATED ANALYSIS

	BROILER STARTER (%)	LAYER PEAK (%)
Protein, % (N × 6.25)	22.50	18.00
ME, kcal/lb	1425	1320
Lysine, %	1.21	0.94
Methionine + cystine, %	0.92	0.71
Ca, %	0.95	3.80
Available P, %	0.48	0.45
Na,%	0.20	0.18
K, %	0.83	0.68
Cl,%	0.25	0.19

aCP = crude protein; ME = metabolizable energy; N = nitrogen.

Dr R. M. Engberg, M. Hammershøj, N. F. Johansen, M. S. Abousekken, S. Steenfeldt & B.B. Jensen (2009) Fermented feed for laying hens: effects on egg production, egg quality, plumage condition and composition and activity of the intestinal microflora, *British Poultry Science*, 50:2, 228–239, DOI: 10.1080/00071660902736722

PELLETS OR MASH?

Mash or crumbles are the preferred forms for chick feed. It is small and their little beaks can easily consume the feed ration. Crumbles are pellets that have been smashed into smaller pieces. Mash has not gone through the pelletizing process. So the question becomes, why pelletize?

Pelletizing is a heat-extrusion process. That heat helps to kill any bacteria that may have come into your mix through infected ingredients. Pellet mills are expensive, and there are very few companies that sell versions that are practical for small flock owners. They can run off of a regular wall outlet but they are bulky, heavy, dusty, and noisy. This would be a good investment for someone who wanted to charge a small fee to pelletize and box up or bag up feed for their neighbors. However, for practicality purposes, the investment would really be for pelleting your own feed foremost.

Free-Choice Feeding

You can employ free-choice feeding at any life stage of your chickens. You may choose to go this route if you don't want to buy bagged feed with the caveat that you will have to buy and store all the ingredients. One research study that offered feed ingredients through free-choice means found that chickens were able to select individual ingredients without detriment to their overall well-being.

PROBIOTICS AND HERBS

There is a growing body of information available on probiotics and herbs for poultry production. These research studies in many cases have found that, depending on the amount of herbs, essential oils, or pre-/probiotics added to the diet, you can have varying responses for growth or even egg production. Oregano oil consistently appears to have a positive effect on controlling foodborne pathogens in the gut and on the final chicken-meat product sold at the grocery store.

So what does that mean for the average backyard chicken owner? Most likely, you are going to need to offer these specialty items as a free-choice option to your chickens. That means you can offer a small container with the herbs of your choosing to your flock and replenish the contents as needed. Probiotics or even prebiotics may be offered as a premix to go into your feed mixer. Keep in mind that most probiotics are a mish-mash of bacteria or other

ingredients and are proprietary or work only sporadically. Companies are constantly trying out new versions when the old ones stop working.

Fresh herbs behave differently than dried herbs or even essential oils. Whichever form you choose to work with, you will need to handle the ingredients carefully. Of course, if you are mixing your own feed and using dried herbs, then you may need to order them in bulk, which means storing them appropriately. Essential oils can be mixed with the fat component of feeds, but some need to be kept in refrigeration or in darkened bottles to prevent them from going bad.

Bear in mind that some essential oils that come from herbs can be caustic in purer forms. Mishandling them can cause skin burns. Working in a poorly ventilated space can, in some cases, affect your respiratory system.

FEEDING EGGSHELLS

Sometimes you'll get a glut of eggs and don't know what to do with them all. If all of your neighbors are full up on eggs, you can certainly freeze your eggs. Crack them into the individual compartments of an ice cube tray and place the tray in the freezer. When they're frozen solid, pop them out of the tray, put them into a plastic bag, and place them back in the freezer. Need just egg whites? Do the same freezing routine, but separate the whites from the yolks before putting them in the ice cube tray.

If you have frozen as many eggs as possible then you may consider feeding them back to the hens. The main reason why many people do this is if their flock appears to be laying eggs with thin eggshells. You can provide oyster shells, free choice, but not every hen takes advantage of that offering.

If you are going to offer eggs or eggshells back to your chickens, then you need to do so with a fair amount of strategy. First of all, the chickens absolutely do not need to know that it is an egg they are eating. That means it needs to be ground into fine pieces. The better option is to hard cook the eggs and then pulse them in a food processor with the shells on. This solid form tends to be a form that the chickens do not recognize, and it can easily be scooped out into feeding pans. Offering this treat to the chickens once or twice a week may help turn things around regarding shell thickness.

Pasture FOR Poultry

Not all pastures grow year-round. Pasture grasses have their seasons, too. That means sometimes a pasture is covered in a layer of snow or the pasture grass nutrients retreat into their roots, only to come back in the spring with new and vigorous growth. So what can you do to provide pasture supplementation year round?

Keeping no waste in mind, you may wish to grow and bale your own pasture grasses. You will need to learn more about the soil type you have and which grasses grow well in your area. You will need to figure out who can come and cut your hay for you because using a scythe is back-breaking work. However, some people with smaller lots use a hand scythe to cut their hay.

You can talk with your County Extension office or maybe even a master gardener to see what grasses or forages grow well in your area. Most states' soil types vary wildly, so you need to talk to a professional. Make an appointment at your local Cooperative Extension, and they will help you.

CUTTING THE HAY

You will need to cut it in the field and leave it to dry. Hay needs to dry (cure) before being baled, and while it is drying after being cut, the weather can wreak havoc on the process. A hay baling machine is needed to bale the hay once it's dried. Sometimes you can hire someone to bale for you for a fee. You can rake it up and store it after it dries in the field, but that is labor intensive and storage takes up a large amount of space. Did you know that they also make mini hay balers? Sure do! But they are expensive and mostly used for decorative straw bales for fall holidays.

Of course, as with any livestock producer, you can buy hay by the bale at a feed store. You will need storage space, however, and a way to keep rodents at bay. Hay makes wonderful housing for rodents that are just trying to make it through winter.

Depending on where you live, it may or may not be easy to find hay. Alfalfa hay is great for chickens but it is among the most expensive. Bales can weigh as much as 100 pounds (45 kg) or more. Luckily there are entrepreneurial people out there who sell "pasture in a bag" and will ship it right to your door. The actual pasture grasses may differ from what you have planted, but at least you will have pasture available to your hens so that you can maintain those wonderful, brightly colored yolks. Yolks are assigned a number from 1 to 16 on the Roche Color Scale. Those are shades of yellow going from very light to a dark orange-yellow.

Fermented Feed

Fermented feed does not have a very precise definition yet. Basically, it is feed that is placed in a bucket and covered with water for a couple days until fermentation begins. It is then fed to the chickens. People who use fermented feed are very interested in alternative or unconventional poultry feeds. The zero-waste takeaway is that chickens like it first thing in the morning, but they will not eat it when it is no longer palatable later in the day.

The claims made by individuals and groups about the wonders of fermented poultry feed are great. However, to date, there has been only one thorough study that looked at laying hens, egg production, gut bacteria, and nutrient utilization as it relates to fermented poultry feed.

That study used Bovans brown-shelled egg layers and had a sufficiently large number of replications to make statistical analysis meaningful. The researchers compared a diet that was dry in mash form with the very same diet that had been fermented for a period of 72 hours, stirring continuously for 30 minutes, 6 times a day. The feed-to-water ratio was 1 part feed to 1.2 parts water.

In this study, birds were provided with more feed than they needed to make sure that they would not run out of food. The birds were fed fermented feed once a day, and the researchers looked at feather condition as well as weight gain and litter quality. We all know that feeding chickens wetted feed is going to result in looser stools. Since the droppings for chickens fed the fermented feed were looser, it also meant that the litter was wetter. It is well understood that wet litter leads to problems with foot pad lesions and can lead to higher ammonia levels in the coop.

One of the more surprising results was that the hens that were eating fermented feed displayed more aggression toward one another and increased mortality. The researchers noted that just because food was always available to them, the birds did not want to eat it after a certain amount of time, and it would dry out a bit in the feed trough. The chickens, although hungry, would not eat the feed after losing interest in it. That likely indicates that the feed, while consumed enthusiastically when it is first served, loses palatability over time. That means that fermented feed does not provide the chickens with the stimulus they need to fulfill their searching and exploration behaviors on a daily basis. Birds would become aggressive and agitated and cannibalize one another in the fermented feed group; a few birds in the study even had to be humanely euthanized. The feather plumage scores of hens fed fermented feed was worse than hens given dry feed.

Researchers also noticed that there was some settling of the feed in the trough, with the heavier minerals going to the bottom. The question remained if the chickens were getting enough minerals to lay eggs, and the answer was "yes." The shell strength and shell thickness, as well as the overall shell weight, increased.

WET FEEDING VERSUS FERMENTED FEED

There is a big difference between wet feeding and giving chickens fermented feed. Wet feeding does wonderful things for chickens and increases the amount of feed that they consume. It also improves how well they are able to utilize the feed and improves their egg mass production. Fermented feed, however, in this particular study was not able to increase feed intake and correspondingly caused hens to delay their start of lay. The weight of eggs from hens on fermented feed was higher, which makes sense since such a large percentage of the egg consists of water. The body weight of hens fed fermented feed was lower than that of hens fed a dry feed but that situation reversed itself significantly by the end of the trial with hens on fermented feed weighing 80 grams (2.8 oz) more.

When you ferment feed, lactic acid bacteria act upon the feed ingredients. As they use the sugars in the feed, the pH decreases from around 6 or 7 to 4 or 5. When chickens eat feed that has a lower pH, such as fermented feed, then the pH level in their intestines lowers to nearly the same levels as the feed. In the study, the pH of the feed lowered, which also translated to lower pH levels in the intestines. The fermentation lowered the pH of the feed to 4.5, whereas dry feed remained near neutral at 6.2. The two feeds did not differ in gross energy, insoluble fiber, starch, fat, calcium, ash, or total phosphorous. The bacteria, through their normal activity, reduced the sugar content by 77 percent in fermented feed. The crude protein of the dry matter was increased by about 3 percent and there was a 24 percent increase in soluble fiber after the fermentation process.

With the limited information that is available, the story that is to be told about fermented chicken feed continues to evolve. The results are a mixed bag for laying hens, but perhaps hold promise. If more research is done to provide clarification in the future, then perhaps a higher recommendation of fermented feed can be made. At this time, wet feed is recommended with the caveat that it is more labor intensive for flock owners to keep things clean. Dry feed is still the easiest to handle and manage for the vast majority of small flock owners.

Treats

Treats are rare and special things that should never be expected or anticipated by the flock. Treats are treats, so that means special occasions or training sessions. Oh yes, you can train chickens, but that would be an entirely different book!

Treats are available as chopped up hot dogs, dried worms, or even whole grains. You will eventually learn what treats are your flock's favorites. If you choose to feed your chickens whole grains as an occasional treat, then be sure to provide your flock with grit to help them grind up the grains. If it is a cold winter and grains are needed for supplemental energy, then provide the chickens with a small container of grains at night before they go to roost and make sure they clean it all up within 15 minutes. That extra energy will be burned at the coldest time of night when the chickens need to stay warm if you do not provide supplemental heat. You will need to wean them off of the whole grains come springtime to prevent an obesity problem! Some experts say encouraging chickens to eat their layer rations provides the same heat-producing effect. Whole grains are often sold as something called scratch or scratch grain. Feeding whole grains is not a balanced diet and should never be the sole source of nutrition for a flock.

Some people like to find toys in which they can place dried worms or whole grains in order to make the chickens work a little harder to get the treat. This also enriches their living environment and helps them achieve their daily exploration and stimulation goals. Hang a few greens in the wire of the coop or put lettuce in a rabbit lettuce feeder for your chickens. Treats are the added stimulation that you occasionally provide when the run or pasture is not as welcoming. Also, anytime a storm is coming, and the chickens need to be "cooped up" for their own safety, it is always a good idea to provide stimulation for them in the form of treats or toys.

Future Feeders

Chicken-feeder designs have not changed much in nearly 100 years. One big leap forward was the use of a lever system to open the feeder when chickens stepped on it. That system kept the feed away from wild birds that want to steal the feed for themselves. This is still in use, but it's pricey and specialized. The majority of feeders are either open trays, with or without a guard on top to prevent walking through it, or a hanging tube feeder.

Innovation has been improved to make designs easier to open and close as well as clean. Plastic makes feeders more affordable and makes it possible for them to be made in a variety of colors to stimulate chickens to eat. Hanging feeders are also a very important improvement over feeders placed on the floor. Keep in mind that feeders should always be kept at the height of the chicken's back to help avoid feed waste.

So that leaves me with where the real innovation needs to be: feed recovery. We have all seen it that one chicken that sits at the feeder and scoops out the feed, knocking it to the floor where it is never eaten. Don't those chickens appreciate all the hard work and money that goes into feeding them? The short answer is, "no."

I am calling upon all you chicken-feeder designers, engineers, and entrepreneurs. Design a feeder for us where feed that is billed out lands somewhere that we can get to. We will happily return the recovered feed to the feeder when we come to the coop for the daily checks. Make it affordable and cleanable so that your revolutionary new feeder can be in every backyard flock owner's coop! Go forth and innovate! I cannot wait to see what you come up with.

Now you have a better grasp of what it takes to offer your chickens a no-waste diet. You are armed with all of this good information to apply as you research what local resources are available. Part of the challenge of chicken keeping is designing a strategy that works for you! Look at the space you have available to you. Then you can decide if you are able to grow and process your own, or if you'll need to visit your local feed store. You may want to start by gleaning waste food from a local school or restaurant.

3

Composting

So you have a mixture of chicken poop and soiled bedding you just cleaned out of your chicken coop. What are you going to do with it? Throwing it in the landfill would be like throwing gold into the ocean. What a waste! You're going to compost it, of course! If you are completely new to composting, then perhaps one of the very best things you can do is to try composting your household waste. It may be something you should consider before keeping chickens.

Take on the process in stages. If you can afford to scale up, then do so gradually so that you can see what you can physically do. It often surprises people how much time it takes to do things right. If you lose interest with a small pile or a pile that is a little bigger, then lesson learned.

One of the major ways to keep going with a zero-waste chicken-keeping lifestyle is to compost on your farm. This is a learning curve and for some of you readers out there, the curve will be steep. You

should absolutely take a composting class at the Cooperative Extension office nearest you that offers such classes. You will need to learn the basics and the rules for your state. For example, in some states, regardless of the size of the flock, chicken manure composters must be covered to avoid runoff into waterways, which causes algae blooms. It is the little details like this that will affect how you tackle this aspect of zero-waste chicken keeping.

The 6 Components
TO COMPOSTING SUCCESS

You need to have a good working knowledge of composting to make it a success. Yes, there are methods of leave it and forget it, but many readers who are really interested in living a zero-waste lifestyle tend to be a bit more active. The golden ratio of carbon to nitrogen that you are aiming for is 3 to 1. That means you are going to need three times as much carbon as you will nitrogen. There is more to consider, so let's head on down the rabbit hole!

1. CARBON

Your compost pile will need to contain about 70 to 75 percent carbon. You are going to be paying much more attention to your compost pile and what you put into it so that you hit this mark.

2. NITROGEN

This is the component that most people think of when they think of composting. But you can seriously ruin, or at least slow down, your composting efforts by adding too much nitrogen into your pile. Your goal is to have this at 25 to 30 percent.

3. MICROORGANISMS

This is really the key to composting success. These tiny organisms do all the heavy lifting for you, no matter what system you choose to use. Where do microorganisms come from? The soil, of course! They are found all around you in the soil. However, it's more likely you can find them in the chicken manure that you are adding to your pile. There are products that you can add to the pile that contain microbial "seed stock," and these products also often speed up the process. These products can be beneficial if you have had a particularly hard season or have waited to the last minute to get started. Hey, it happens to the best of us!

> An example of a slow or holding-style composter. Fill up one side and it will be ready for use the next year.

4. WATER

The moisture in your pile will need to be uniform throughout. Not dripping wet, mind you, or you can wind up polluting your local waterways! More than likely, you never thought your little compost pile could be considered a detriment to your environment, but you will need to manage your pile so you do not add to the troubles of your local waterways. As you add layers to the pile, spray the pile with water. If a storm is on the horizon, then something as simple as a tarp over the top will prevent your compost pile from getting drowned out.

5. OXYGEN

Although there is a great variety of microorganisms that are living in a compost pile, it is not necessarily meant to be an anaerobic, or oxygen-free, environment. Air, or oxygen, gets worked into the pile when you turn it. The size of the materials in the pile also allows pockets of air to work their microbial magic. Turning "fluffs" the pile and retards the growth of certain non-oxygen-loving microbes.

6. HEAT

Some like it hot and that includes your compost pile. As microbes do their beneficial work in your compost pile, they create heat. There are gradations in heat as a pile moves from new to fully composted. The heat from a turned pile will kill most weed seeds and bad bacteria.

Find Your Style: CHOOSE THE RIGHT METHOD FOR YOU

Composting can be hands off or hands on, and this depends on your farm as well as on you. You can use a holding method, a turning method, or a heap method, and then there is sheet composting. Let's take a look at each type individually and see what may be the most attractive to you. (Note: Food scraps or mortality loss [dead chickens] are really not supposed to go into a compost heap because of the possibility of attracting unwanted scavengers.)

HOLDING METHOD

This is generally a slower method and is a great way to get compost after a year or two. You do not turn the pile; rather, you just add to the top and sometimes spray the pile with water. If you are just using a one-bin system, then oftentimes it is just a holding unit. Oxygen is not worked into the pile so it tends not to get as hot, so do not expect to see steam rising off the pile like you would for a turning method.

You do not need much space, especially if you are just working with a small six-bird flock. And if you set up two of these piles, then you can have a pile almost ready to go after a year and start adding to the second slot in year two.

The stages of decomposition vary as you move upward toward the top of the pile. The more composted material is at the bottom waiting for you. You can attempt to better aerate the pile by using a shovel to make holes in it or adding PVC pipe (at least 1-inch [2.5 cm] diameter) with holes randomly drilled in it to the pile. PVC not your thing? Then how about a woven stack of twigs interspersed throughout the pile? You can also put your pile on a pallet or even a plastic aeration mat, if you can find one.

TURNING METHOD

This system is designed to compost faster. You turn, or aerate, the pile using either a pitchfork or maybe even the front loader on a tractor. Some people even use their chickens as aerators because they dig down into the piles.

If you choose to use a bin that spins, it will need to be on a stand of some sort. The moisture content of the bin will need to be maintained as much as if you'd chosen to use a pile. You should, without a doubt, handle a turning bin to see if you are physically capable of making it work. Those with back problems may opt for the holding method over the turning method.

You can plan on making compost in about 2 months but turning will need to be done regularly. That means every 5 to 10 days. Temperatures get higher. Piles do have the possibility of getting too hot and catching their surrounding structures on fire. Your goal is to have a pile reach above 130°F (54°C). Not sure if you are reaching a temperature that is high enough? Well, a compost thermometer is just the tool to help you. Essentially it is a very long probe that you put into the middle of your compost pile to determine the pile's temperature.

HEAP METHOD

No structure is needed for this method! Nope, no expenditures there. Rather, what you are doing is creating a large pile that is 5 feet (1.5 meters) wide by 3 feet (91 cm) tall. Make the pile as long as you want it to be. Why 5 feet (1.5 meters) wide? Well, it will help you retain the heat for the composting process. Space is the biggest issue here because you can add to one end of the heap while the microbes are finishing up their work at the other end. Turn the pile or do not turn it; this is completely up to you.

SHEET COMPOSTING

This method is more of a soil-building measure than a traditional compost pile. You will spread a thin layer of waste material over the soil that you plan to till. You will put down this layer at a 2- to 4-inch (5 to 10 cm) thickness. Work that material into the soil in the fall and it will decompose over the winter. If your waste material contains more carbon than nitrogen, then you may accidentally end up "pulling" more nitrogen from the soil to aid in the decomposition of the carbon material.

DIY Composting

There are as many different ways of building a compost pile as there are ways of building a coop. Here are some general guidelines to help you avoid rookie mistakes.

If your composting unit is made of wood, then avoid using treated wood. What are the components of treated wood that you will need to watch out for? Chromated copper arsenate (CCA), creosote, and penta. So as tempting as it is to get old telephone poles, be sure to ask with what chemicals they may have been treated. Some of the chemicals we mentioned are toxic compounds and can be harmful to humans and pets.

There are some woods that are naturally resistant to decay. It is going to be more expensive but get some cedar planks or grow your own cedar trees (warning, this is going to take a while!). Untreated pine will also do the trick, but it will decay and need to be replaced after a few years.

Wire or metal does not decay but it can, at times, be more expensive to purchase initially. If well cared for, wire or metal allows adequate aeration and will last a decade or perhaps more. You can even make your composter round, given the flexibility of wire.

For a more permanent solution for a composting system, use stone or concrete. Concrete cinder blocks also make sturdy holding containers. Even if your bin sits directly on the ground, you can use damaged concrete road dividers or cinder blocks to make walls for your composter. You can easily turn the pile in a composter that is large enough. The nice part about using more-permanent materials is that they can be sturdy enough to handle material in larger quantities or turning via a front loader.

CHOOSING WHAT TO ADD TO YOUR COMPOST PILE

What goes into a compost pile? Well, absolutely no plastics or any waxed paper products. Send those items to the recycler.

Remember the 3-to-1 carbon-to-nitrogen ratio? You want most of the compost pile to be carbon so that means poultry bedding, shavings, straw, dead leaves, wood chips, and garden waste. If you are going to use yard waste, then the smaller you chop it, the better off you will be in breaking it down in your pile. Green materials are things such as food waste, grass clippings, vegetable waste, and fruit. Avoid things such as meat or dairy products as they are proteins and can be disruptive to the composting process.

So, let's play a game: Is it brown or is it green?

Eggshells do not break down in a compost pile. If they're just thrown in the pile, they will stay "as is" for years. They have even found eggshells in archeological dig sites!

So how do you make a profit from this waste product? Easy! Wash out the shells, let them air dry overnight, and grind them to a fine dust in a food processor. You can add this dust to the compost pile and you will enrich the final product. We also recommend that you sell it by the bag to folks who need the added calcium in their gardens. A few minutes of extra work for you can translate to a few extra dollars in the bank.

Brown	Green
Sawdust	Chicken droppings
Shavings	Veggies from the garden
Moldy chicken feed	Fallen fruit=green

Empty Eggshells = Neither, they add calcium

FINISHED COMPOST: WHAT NOW?

Compost is, of course, a hot commodity around a farm. You can use it in your garden or anywhere that you plan to till the soil. Rather than buying a bag of compost or fertilizer to add to your lawn, go ahead and spread a ¼-inch (6.4 mm) layer of your own compost on your grass. Your flowerbeds and container gardens will heartily reward you with the addition of more blooms if you use compost in the potting mix. Compost is also a good mulch material for some plants in your vegetable garden.

If you compost more than you need in your own garden, you may want to give your neighbors a couple of bags of it. However, you could make a few extra dollars by selling some of your compost. Let the benefits of your compost spread by word of mouth. Perhaps take out an advertisement in the paper or post information on how to reach you in a neighborhood paper or online marketplace. Let master gardeners know and they will tell others. The best way to sell your compost is by the bag, and if you have leftover feed bags sitting around, then by all means fill them up!

Composting is a process, so patience is required. Now that you know what you are doing, I encourage you to go out and try your hand. Your garden and neighbors will thank you!

Composting Sales

You might find good local compost through farmer's markets, online sources (like Craigslist or Facebook Marketplace), or agricultural advertisement papers like *Farmer's Market Bulletin*. For some people, a community that communicates well often finds those people who are excellent at composting. Then everyone is knocking at their door to buy compost! This can be you. Compost from your chickens can be sold by the bag. Where are you going to get bags? Well, if you bought chicken feed by the 50-pound (22.7 kg) bag, then you have plenty of bags. Get the word out by putting a sign at the end of the driveway or put fliers up at local nurseries. Refill those feed bags with compost and sell, sell, sell!

How to Compost in 10 Easy Steps

The best method explained for backyard chicken owners was a publication by the University of Georgia's Cooperative Extension system. They recommend covering their compost and using a three-bin turning system. An excerpt from their guide follows:

STEP 1

Rake litter from poultry housing area on a weekly basis if you have six or more chickens. If you have five or fewer birds, you can probably get by with cleaning out litter every two weeks or so. More frequent cleaning minimizes insect, odor, and pest problems.

STEP 2

Add poultry litter to the first bin. Because poultry manure by itself is wetter and higher in nitrogen than poultry litter, you may need to mix in materials such as leaves, straw, wood shavings, or wood chips to balance nitrogen with carbon, add bulk for air circulation, and absorb excess moisture.

STEP 3

Cover the compost pile with a solid roof or tarp.

STEP 4

Over the next few weeks, continue adding litter and other compostable brown and green materials until the bin is full. Mix the ingredients well and spray with water as necessary to achieve even moisture throughout the pile. Cover the pile between additions.

STEP 5

When the first bin is full, cover it and allow it to compost undisturbed for two weeks. Monitor temperature with a compost thermometer. Aim for an internal temperature between 130°F to 150°F (54°C to 66°C).

STEP 6

After the pile has composted for two weeks, turn it into the second bin. Cover the pile and let it compost for several more weeks.

STEP 7

Repeat steps 2 through 5, piling fresh litter into the first bin that you had emptied previously.

STEP 8

When the pile in the second bin has composted for several weeks, turn it from the second bin into the third bin to allow it to mature for several more weeks.

STEP 9

Turn the material from the first bin into the second bin for a second heating cycle.

STEP 10

Repeat steps 2 through 9.

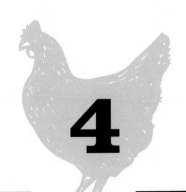

4
Gardening
WITH Chickens

There is a perception that chickens are the be-all, end-all for gardening. They are certainly fun to spend time with. They are natural bug hunters and manure depositors. They till the earth with their feet and toenails. But there is more to this equation than you think.

When most people think about gardens, they think of the plants and what they can harvest from them. It is a nice thought, but it is really the last step in the process. Step one is choosing the location of your garden. Step two is planning out the space you need for each of those plants. Step three is soil preparation.

And did you know that the chickens can actually help work the garden? Since waste expends resources to no purpose, profiting from additional work our chickens can do is the opposite of waste. Now, there are some important factors to know before you just throw you chickens in the garden.

During the dead of winter, seed catalogues arrive that both torment and thrill gardeners of all ages. Thumbing through the pages, you dream about new and exciting plants. Feed your family, preserve some for next winter, give the extra to a local food pantry. (You should absolutely plant extra rows to help end hunger in your community.)

You know that gardening can be time consuming, but you love the experience. If this is new to you, then get ready to be hooked! Grab yourself a good pair of gloves, a hat, a hose, and a trowel; put on a little sunscreen and a long-sleeved shirt; and borrow a tiller from a neighbor so that your first time gardening is a great one.

You have your land. You have a flock. You know which plants, vegetables, and fruits your family like to eat. You have a green thumb! How can you make all this work well together? Let's talk about a way to put all of the compost you have been creating with your flock to work!

In this chapter, we'll explain how to use chickens in your garden efficiently and give the scientific explanation behind it.

CHOOSING THE Location FOR YOUR Garden

Not all spots on your property are particularly well suited for gardening. You will need to look at the lay of the ground. Do not put a garden in a low spot where mud and rain runoff are going to be issues. Shade is also not ideal; most garden plants prefer full sun for at least 6 to 8 hours a day. And get ready to have a little extra space. There is no doubt that something extra will tempt you in the way of seeds or started plants when you are out and about shopping locally.

You are going to think you have a perfect and picturesque spot in your yard for a garden, but you need to think about space. This is where planning and ordering seeds in advance will help you. There is good information in most catalogues to help you plan. Then get out some graph paper and plan the garden out. Each $^1/_4$ inch (6.4 mm) on graph paper can translate to 1 foot (30 cm) of yard space. Go outside and measure; do you really have enough space to do what you want as far as a garden? If not, then go inside and rethink some of those details.

Gardens do not do well if they are located in overworked, tired, or diseased soils. So you may need to be flexible about moving your garden from year to year. Also, certain plants need to be rotated within a plot to avoid picking up certain known diseases. For example, tomatoes, a member of the nightshade family, can potentially develop disease if they're put into the same plot where you planted tomatoes or another member of the nightshade family last year.

PLANNING THE SPACE YOU NEED FOR YOUR PLANTS

Pay attention to seed packets after you have chosen which seeds are making a debut in your garden. These packets contain a ton of wonderful information that not only is specific to the plant but also to the variety that you have chosen. This information includes planting depth and spacing. If you plant your seeds too close together, then you will get overcrowding. You will then need to thin the seedlings shortly after they come up. Of course, this will be necessary with some really small seeds like carrots and radishes.

Transplants are a little easier. But do not toss out the labels on those, either. They have two handy purposes. First, they have the same good information on them with regard to spacing as the seed packets have. If you overcrowd plants in the garden, they may fail to produce adequately or may even invite plant diseases because the plants are stressed. Secondly, you can put those labels in the ground to mark what you planted in that particular row.

PLANNING A SPACE FOR YOUR CHICKENS TO BATHE

Chickens are not like other birds. They do not bathe in water sources like your local mockingbird or finch. They bathe and clean their feathers using dirt. That means they will actively seek out loose, friable soil and take a bath in it. With this information in mind, you will want to plan and provide an appropriate bathing space or your chickens may choose an inopportune location.

Much in the same way that some dog owners plan outdoor spaces with their dog's habits in mind, you, too, can plan a space for dust bathing (see more on page 106). You can use an old sandbox or set aside a space in the yard in which you place sand for the

chickens. They will enjoy scratching around in this loose material. You will need to pick up droppings in this area. Otherwise, over time, chickens will "enrich" your dust bathing area with manure and plants can gain a foothold. A space that is 9 square feet (0.8 square meters) is small, but it will suffice for a couple chickens to dust bathe together. Make sure this part of the yard drains well so that your chickens can use it year round.

Soil Preparation

This step takes time! Not all soils are up to the task of feeding your plants and, by proxy, you and your family. Good, rich soil takes work. Your compost is going to come in handy, indeed! After you do a soil test and speak with professionals about the results, then you can formulate a plan of action with regard to adding compost to your garden soil.

In every county of the United States, there is a Cooperative Extension office. They provide this testing service or will send a testing sample elsewhere for a small fee. When you get the results back, you can talk to someone in the Extension office and they can help you. Also, master gardeners, many of whom work through the local Cooperative Extension office, can help interpret the results.

TILLING THE GROUND

How do you plan to till the ground? Your chickens are not going to be able to break sod for you, mighty as those beaks and toenails may be. You can certainly allow your chickens to follow behind you as you fire up a tiller or hire a team and plow. A tiller, however, tends to be a rather novel and scary-sounding behemoth in your yard. It may be avoided like the plague by the members of your flock. Unless your chickens are some of the bravest and most intrepid out there, then you should introduce them to your freshly tilled garden very quickly after you have put away that demon tiller.

The cacophony of sounds that are likely to come from your flock once you have released them on freshly tilled ground is certainly going to be worthy of a YouTube video. Let your chickens out, and they will eat tender weeds and short grasses. They will demolish earthworms and grubs, spiders, and grasshoppers of all shapes and sizes.

Herein lies a risk along with the reward. You may enjoy watching your chickens do the work for which you might otherwise need chemicals, but you also need to understand certain life cycles. Some internal parasites of birds, including those specific to chickens, spend part of their life in some of those earthworms and insects that you just tilled up. Quarterly, you will need to do a fecal check on your chickens to see if they have picked up any of these nasty perpetrators of intestinal havoc.

The large roundworm and the tapeworm each have unique skills. They can complete their life cycle either by direct or indirect means. A chicken can become infected either by eating the droppings of an infected bird or they can consume an intermediate host. Intermediate hosts, quite conveniently,

include some of chickens' favorite garden treats. Those are earthworms, snails, and insects such as grasshoppers or spiders. The moral of the story is this: If an insect, snail, or earthworm eats worm eggs deposited from an infected bird, and then your chicken eats that intermediate host, then the eggs can hatch inside the chicken's gut and attach to the intestinal lining.

Internal parasites steal nutrients from their host and take up valuable space in the lining of the intestines. You can do a fecal flotation test through your veterinarian on a quarterly basis and the results can help you determine when it is time to perform a strategic worming effort! There is nothing like watching your chickens go to town on some bugs in the garden, but you will need to do your part to make sure that their insides are working smoothly.

RAW MANURE IN YOUR GARDEN

Some people like the idea of having chickens walking up and down their garden rows working on their behalf. However, chickens produce up to $^1/_4$ pound (113 g) of manure a day per bird, albeit only a small amount of this manure is deposited while the chicken is in the garden! Regardless, raw manure is raw manure, so we need to look at this issue carefully in order to understand the risks.

Raw manure, just like compost, is a soil amendment. You should not apply soil amendments during the growing season. Why? Well, foodborne pathogens have really awesome skills! They can be in raw manure or improperly composted compost and be taken up in the roots of a plant. Once absorbed by a plant's root system, they can also make their way into other parts—translate that to edible parts—of a plant. Now, this is not a problem if you properly cook your food and do not have cross-contamination in the kitchen.

As we all know, not all plants that we humans eat are cooked when we consume them. That does not mean that we do not wash or rinse them prior to serving. This also applies to your herb garden. Herbs, vegetables, and fruits should be washed to remove dust, dirt, droppings from all kinds of critters, and any foodborne pathogens.

Washing is a good food-safety step, but it will not wash off foodborne pathogens that have made their way into the interior of the plant. What on earth do we do?! We certainly do not want to have to cook all our veggies. That defeats the purpose of a fresh salad!

Well, we are certainly not the first, nor will we be the last, to tackle this issue. There are some very fine scientists who have been working hard on your behalf to hash out some of these details. Certain organizations that work with growers have mandated that you cannot harvest leafy greens from soil that has had raw manure applied to it unless a year has passed from the time of application and the harvest. If you belong to certain organizations in order to get their seal of approval, then you have to follow their rules. Two of these are Certified Organic and Cornucopia Institute, but there are many such organizations around the world.

Sounds extreme, right? Well, think back to those outbreaks a few years back in which many people were sickened from bacteria obtained from leafy

greens. Some of those individuals died. Salads are really tasty, but they're not worth dying for! One of the more frustrating things is that much in the way of food-safety information used to be covered in home economic courses offered in many junior high, middle, or high schools across the country. Now this information seems to be lost to common sense and a blame game ensues.

There are other recommendations that parse out the details a little further. The rules from the National Organic Program (these are called the NOP Standards and they can be found at www.ams.usda.gov/about-ams/programs-offices/national-organic-program) differentiate between plants that have edible portions that come into contact with the soil and plants that do not. If you have plants that have edible portions that come into contact with the soil (carrots and leeks are good examples), then it is recommended that you not apply soil amendments for 120 days ahead of your harvest. That is 4 months, and this period applies regardless of whether you are starting from seed or putting in transplants. Some of you may think this still sounds extreme, but wait, there's more!

We discussed plants that come into contact with the soil. There are plants that do not have edible portions that contact the soil. For example, we have corn, sunflowers, and several bush and climbing bean species. It is recommended that you not apply soil amendments for 90 days ahead of your harvest, which translates to just 3 months.

Let's take a step back here for a moment and really think about how to manage the space in which the family garden is going to grow. Now for some tough questions. Do you have enough space to let your garden sit after letting your chickens go to town on it at the end of the growing season? That may mean you let it sit for a year if you plan on growing strictly leafy greens. You recall you will need to rotate your garden from year to year to allow the soil to recover and break the life cycle of certain soilborne plant diseases.

In the middle of winter, when most of us are thumbing through both plant and chick catalogues, just dreaming of the spring and summer to come, be sure to look carefully at a plant's growth periods. In your catalogues, you will find the average number of days for each plant until you can harvest your food. You may need to pull your chickens off of the garden plot sooner than you expected in order to grow all the things your family likes to eat. Or, if the weather lets you, you may have to pull out the demon tiller and till the ground sooner than you normally do. That may also mean that you let your chickens go to town on the insects for a shorter number of days than they have in the past.

You can still let your chickens run in the garden, but a potential solution to the raw manure situation may seem more ridiculous than the risk of foodborne illness. Go ahead and till that soil and let the chickens on that plot. Put down some of your lovely, black, rich compost from the coop and let your chickens till that in, too! Let your chickens loose in the garden to hunt for insects and grubs. All the while, your chickens can be wearing diapers! Not exactly zero waste, but if they are made from a washable material, you can just drop them in the washing machine. There are chicken diapers available made by fellow chicken keepers, and they can be ordered to fit chickens of almost any size and shape.

COMPOST TEA

Sometimes people think about using a compost tea in their garden. It is an interesting way to get mileage out of your compost. This is doubly true if you are unable to get your chickens in the garden to till in the compost on your behalf.

The definition of compost tea is a water extract of compost produced to transfer microbial biomass, fine-particulate organic matter, and soluble chemical components into an aqueous phase, intending to maintain or increase the living, beneficial micro-organisms extracted from the compost. According to the recommendations of the National Organic Program, a compost tea should be made with potable water and all equipment should be thoroughly sanitized beforehand using NOP allowable materials from their list. Compost used to make the tea should come from manure or feedstock sources, otherwise bacteria levels may be too high. Compost tea that you make at home without additives can be applied to your garden without restriction, which means you can add it whenever you like. If you do buy and use any of the numerous additives, then the NOP asks its participants to do testing before application to make sure it is safe. You may wish to do the same.

If you used an additive and tested the final compost tea product, and you plan to use it in your vegetable garden, then you need to follow the 90/120-day preharvest recommendations. If you are just using the compost tea made with additives on ornamental plants or grains, then you do not need to follow the 90/120-day preharvest recommendations. If you make a compost tea from raw manure, then you should definitely use the 90/120-day preharvest recommendations. You should never apply a compost tea made from raw manure on the edible parts of a plant in your vegetable garden.

THE NINA, THE PINTA, and the Silver-Laced Wyandotte?

Like the great explorers of the eighteenth and nineteenth centuries, your chickens will go forth into your garden and seek out all that they find to be interesting and edible! Think about it: Botanists and explorers sought new places and more information that they could share with others. Those newcomers to many parts of the world also sought great wealth and perhaps wanted to find El Dorado, but have no fear that your chickens will seek out such material things.

Your chickens, when let into your garden, will see it both as a new playground and an old familiar couch all at the same time. They have their favorite spots in which to hang out, but at the same time they appreciate the details. They do not see the world in the same way that we do. They can see different wavelengths. That means that they can take in and see changes to the plants that we may miss. We miss the parts of the plant that are in wavelengths that we lowly humans are unable to see with our meager vision. Just imagine what birds and bees see when they fly around.

The visual acuity of a chicken can exceed that of our own in some cases. They can go through a patch of plants and catch sight of a bug that we would never notice. And then the bug may be picked up deftly and accurately from whence it sat with hardly a disturbance to the plant it was hiding on.

Flight may not provide a sufficient measure of safety for an insect either. Chickens do better tracking movement than you may think, and they can track flying insects well. The challenge presented by an insect in flight is just what the doctor ordered with regard to keeping a chicken's mind occupied and a chicken's crop full.

The majority of a chicken's day is spent in exploration. That is what is hardwired into their repertoire of chicken behavior. That means about 60 percent of their day is spent looking around at things, seeking food items, and finding fresh water. What do they do with the rest of their day? Well, a lot! Dust bathing, forming alliances, eating and drinking, preening, asserting dominance, and mating or

caring for chicks are all things that stimulate the brain and occupy a chicken's mind. This can all be done in or around your garden space.

So in the exploration of a chicken, what is it that they *do*? Well, chickens have great eyesight and can see a lot of detail down close to the ground in your garden. They take it all in. They also do not have hands. So that means when a chicken wants to investigate something further, it has only two options. Either things are picked up with its beak and assessed for their "eatability" or things are moved aside with its toes through scratching. Sometimes nonfood items, such as small stones or grit, are consumed to aid with the grinding of food in the gizzard.

So what about the whole deal using chickens for tick control? Well, the research just doesn't exist. It is a popular internet myth. The studies that were done years ago were done with guinea fowl, not chickens. The studies were also not easily replicated and so the research was not repeated. No organizations have funded further study of that small, localized study.

Small, tender plants are consumed, as well as insects and seeds. The beak design of a chicken is that of a mainly granivorous bird, but the gastrointestinal tract indicates that they are omnivores. That means they can eat plants to some degree, but also grind up seeds to obtain nutrients. Meat is not out the picture, however, and any unfortunate mammal, reptile, or even a tiny bird fallen from a nest may be consumed with gusto! Let's go back to the plants for a moment.

Any small, tender plant can be eaten by a chicken. That means if you let your chickens out into the garden before the plants are large enough, then your chickens may just end up eating the new growth. They may not even thank you for it!

At what stage should you let your chickens out into the garden? Well, for some plants, the answer would be "never." That means that certain plants, such as lettuces, are fair game all the time unless they grow tall as they go to seed. The best bet for garden success is to keep your chickens away from any sprouting seeds. Also, young transplants are easily crushed. Other plants need to be large enough to handle the chickens leaning or stepping on them. That is usually about 6 to 8 weeks after planting transplants. That does not mean that the chickens will not peck at and perhaps destroy the leaves or even the edible portions. Tomatoes are usually red, and red is a very, very stimulating color to chickens. Do not be surprised if you find that lots of your tomatoes contain peck marks! Chickens can be trained to prefer and to eat certain weeds. That

means you will need to pull said weeds and feed them to your chickens so that they learn to identify and like the taste of them. That does not mean, however, that your chickens will not suddenly switch to other plants in your garden, possibly your vegetable plants, should they discover that they taste better.

In a preliminary study done at a university, chickens were placed in a pen with pole lima beans, which grow on long vines, mostly above the heads of the chickens. The plants did well after interactions with chickens and the chickens did eat a few of the leaves, but not many. Insect pressure on the plants was not lessened a great deal by keeping chickens with the plants.

The chickens did well in the middle of summer by retreating under the vines to cool down in the middle of the day. These taller plants also had the added benefit of providing shade between the rows, and the chickens took advantage of that aspect. Chickens evaded aerial predators by retreating under the vines or entering the coop. It was also hoped that the chickens would keep weeds between the rows down, which they did in the first 2 weeks. Then, as the grasses and weeds grew taller and thicker, the chickens did not find them palatable.

The next study looked at keeping chickens in with bush beans. The chickens did significant damage to the plants. The insect pressure was affected simply because there were fewer plants in the pens with the chickens. Future research will hopefully give us more information on when chickens should be put in with lower-growing plants like bush beans over that of climbing or taller plants.

The Slings and Arrows of Bugs and Plants

Negative interactions in the garden are most certainly a possibility. Not all insects respond well to being pecked. Bites and stings can happen. Chickens do not necessarily recognize which insects should be eaten; rather, they take the approach of "if they could be eaten." Some caterpillars have wonderful defense mechanisms. Some spiders have a poisonous bite. Bees and wasps sting.

Examine your chickens daily after they go to roost in the coop at the end of the day. Look for swelling or redness, especially around the face. Stings or bites in this part of the body may cause swelling that can affect the airway. Listen to your chickens at night to see if each one is breathing normally or if you detect a flock member with a slight whistle to its breathing. Tail bobbing is also an indication that a chicken is working harder than normal to breathe.

Not all plants are meant to be eaten, either. Many of the same plants that we warn humans away from also apply to chickens. Decorative plants such as nightshade, which produce berries, or long-leaved blooming plants such as oleander, should not be consumed. Be wary of mushrooms that pop up in the lawn during certain times of the year. Before releasing your flock in your yard, including the garden, be sure that you do not have any poisonous plants to which they can gain access.

AT **Year's End**

When the garden is done and looking extra sad for all of its effort in the year, it will be time to send in the beaked and taloned warriors of the yard. They will work with you as you pull up weeds, yank out spent plants and vines, and remove any groundcover you put down for weed control. Oh, the joy that is found working alongside your chickens in the garden.

You can let plants lie in the garden before moving them to the compost pile or maybe you want everything all cleaned up in one weekend. Either way, your flock will help you with insect control and by eating up any plants that were perhaps too small or irregular to harvest. And if you should find a stray frog or lizard, then it is certainly game on for the chickens!

Chickens can make excellent gleaners. Gleaners come along after a harvest and pick up food that was otherwise considered left behind. This applies to gardens, vineyards, and orchards. In fact, chickens do really quite well in all these settings. Your biggest responsibility as a flock owner is to pick up any rotten vegetation before your chickens get to it. No matter the plant material, rot leads to bacterial and fungal growth, both of which can be detrimental to chickens. Some of the same toxins that cause humans issues from rotten food can also affect your chickens.

Orchards have fallen fruit or nuts, though the nuts may not be accessible to your chickens unless stepped upon. But the orchard and vineyard have added benefits. They provide outstanding cover from aerial predators. And as long as your chickens do not have clipped wings, they can fly up to the branches and hunt for insects in the summer months. Those menacing stinkbugs or Japanese beetles are certainly fodder for hungry beaks. There have also been instances where poultry have been moved through vineyards periodically and, if the grass is not too high, then they will mow it all back down.

Chickens are nonruminants. That means they cannot eat grasses and plant material to access the nutrients within as effectively as a ruminant animal. Cattle, sheep, and goats are excellent examples of ruminants. The digestive system of a ruminant is designed to break down the protective layers of grass and plants and access nutrients. That does not mean that plants are defenseless against grazing animals, including ruminants. They carry within them antinutritive factors designed to deter these plant predators.

Legumes are digestible for chickens, as are short, tender new growth grasses. Legumes, you say? Yes, indeed! Clover is your friend and a chicken's BFF! Your local bees will thank you, too! White clover is a terrific forage for chickens. It stays low to the ground and does not get overly rangy even if it is not foraged upon for a couple of weeks. Alfalfa is another good choice, although it will need to be cut

and baled every 45 days or so. Vetch is another good choice, but it is not preferred by all farms.

So how do you maximize pasture and grasses for the benefit of your chicken flock? Well, just as was mentioned, chickens prefer young and tender grasses. If it gets much above 3 inches (7.6 cm), they tend to start ignoring it. If grasses get to about 6 inches (15 cm), you start to get into grasses that are too long. Long grasses can lead to crop-impaction issues. If you can plant and mow your pasture in a staggered manner, then you will always have fresh pasture to move your chickens onto when the old one is spent.

5

Housing Chickens

There are no perfect chicken coops. Sorry, but that is the truth. We encourage you to see the coop as an open door for your creativity, but there are coop design principles that you need to give some thought to before you hammer the first nail. The same goes for brooding chicks. We know, we know … you just want to get started with your biddies, so why can't anyone ever provide you with a simple coop recipe for success?!

When talking about housing chicks and chickens, no waste will look different for different people. Some people may not have the skill set to design and build their own chicken housing, but they do have money to spend on a manufactured brooder or a made-to-order coop. In this situation, it might be a waste of time, resources, and energy to try to build something when this is not your area of expertise. On the opposite end of the spectrum, other people may be creative and have the engineering skills that would help them to successfully build an appropriate home for their flock with minimal impact on their physical and financial resources. Think about where you fall within this spectrum and how you can provide your flock with adequate

housing in the least depleting way. Remember, pretty is not always convenient, and convenient is not always pretty, but it can be both with more effort. Think about what is most important to you and look for or design housing that meets those needs. In this chapter, you'll learn how to repurpose materials, how to seal surfaces, remove hazards, provide adequate space for each bird, and think seriously about your design's "cleanability."

There is much to think about with regard to housing and brooding chickens. This chapter encourages you to take the time to really look at your outdoor spaces and plan ahead for practicality, because mistakes can be costly for your flock. After that, the world is your oyster, er, egg basket.

The Big Shift

When it comes to housing chickens, you have some important decisions to make. Once you make the decision to raise chickens, you'll begin by choosing a brooder, which is where your chicks will live until they're old enough to make the move to a regular coop. Your options for a brooder range from a cardboard box to a galvanized metal stacking brooder, each of which has its own features. You may even opt to put an old, unused refrigerator to work as a brooder in your garage or to take advantage of a little-used bathroom in your house and use a bathtub as your brooder.

Again, the best no-waste brooder will look different based on each person's preferences. For some, having the chicks nearby in a secondary bathroom will be important to saving time. People who don't brood chicks repeatedly may choose just to reuse the box their new microwave came in. If you frequently brood chicks, you may be more interested in having something sturdier that you can use over and over again and will withstand the test of time.

Whichever option you choose, you'll also need to decide how you will heat the brooder so that your chicks will be warm and comfortable. This will be the most expensive part of brooding if you choose the least expensive option, so you'll definitely want to make some savvy choices about how you will keep your chicks warm and comfy.

> These chicks, perhaps like yours, may be destined to live comfortably in a no-waste chicken coop.

Before your chicks are big enough to move out of the brooder and into more spacious quarters (a coop), you'll need to choose the type of coop, all while focusing on your goal of zero-waste chicken keeping. A coop can be as small as a repurposed doghouse (great for two or three chickens) or as large as a trailer, wagon, or RV, any of which can be cleaned out and repurposed as a coop. These and other options will be outlined in this chapter, along with information about nest boxes, perches, feeders, and waterers. You'll also become an expert on dust bathing, a practice that chickens engage in naturally to keep their feathers clean and one that you can facilitate in a zero-waste method to avoid the development of craters in your coop. Read on to learn all about the housing options for your flock.

Brooder
FOR THE LONG TERM

GENERAL DESIGN AIMS

Brooders are where you are going to raise chicks until they are old enough to move outside into your regular coop. There are several reasons to brood chicks in a separate container from that of the coop that will be their final destination. Seasons, design, convenience for the human caretaker, and the overall experience are some of the reasons that can influence your decisions.

If you are starting your flock in the fall, then the weather might be too cold for brooding outdoors without supplemental heat. Perhaps you have a cold snap and need to button up the coop to make it warmer while trying to brood chicks. While that may work for some of you, for others it may prove too much of a challenge in controlling drafts.

Not everyone can afford a nice galvanized metal stacking brooder like those that are sold online by some of the backyard poultry-equipment companies. However, they are incredibly invaluable when weather takes you by surprise. Brooding indoors in a space that does not get much foot traffic or have a problem with drafts is ideal, especially if this is

your first time working with chicks. Keep in mind that brooding is a dusty business and that the space will accumulate a great deal of dust kicked up from shavings and dropped natal down. Natal down is the layer of first feathers on a chick. They are pushed out and discarded by their larger feathers that grow in after the chicks are a few weeks older.

There is nothing worse than being in the middle of brooding and not having the process go as planned. Convenience makes things easier, and it also helps you catch mistakes or mishaps earlier if you are spending more time with the brooder. Having the brooder in the home, garage, or a secondary outdoor structure also means you can keep potential predators at bay. These locations also make it easier to view the chicks, which can be a tremendous learning opportunity for children and adults alike. The bonding over chicken raising is fun! Everybody likes learning more about the food they eat and how it is raised.

WHAT SHOULD A BROODER BE MADE OF?

There are plenty of zero-waste or low-waste ways of brooding chicks. The aim is to make it easy to clean weekly. Plastic, metal, and painted wood are your friends when it comes to brooders.

A cardboard box as a brooder may seem like a low-waste solution since you are repurposing an object. However, should it get wet from a spilled or dropped waterer, then it will disintegrate and need to be replaced. Also, the space needed for each chick increases as it grows each week. If you don't start with a large-enough box, you'll be in search of successively larger boxes each time the chicks run out of room, which defeats your low-waste goal. We suggest that you try something different and more permanent. The plastic bin and wooden box are zero-waste methods as long as you keep your materials in good shape. They can be used for multiple brooding experiences and will last you a dozen years or more!

Plastic bins are affordable and do double duty as storage containers. A simple window screen over the top can keep pets from harming your birds and prevent older chicks from flying out. In some cases, depending on the screen size, you may also be able to keep some dust at bay! Since the bins are made of plastic, they are super easy to clean. Depending on the size of the bin and number of birds you have, you will be cleaning your brooder out at least once a week and possibly more often as the birds grow. It is nice to know that you can simply move your chicks into the next size up in the plastic bins and just dump the shavings and waste into the compost pile. After that, you can hose out the bin or power wash it using soap or disinfectant and let it air dry. Sunlight and fresh air do as much good in reducing bacterial counts as some disinfectants on certain surface materials.

> ❯ Chicks appreciate a well-maintained brooder. It is up to you to see to their needs in the brooder of your choosing.

> Brooders are social spaces and become crowded as chicks age. Over time, get ready to provide them with more space.

Alternatively, a large wooden box works nicely. However, the larger it is, then the harder it will be to cover it with something so that your chicks will not fly out. Additionally, the wood will need to be painted so that the surface can be scrubbed clean on a weekly basis during brooding. The larger the box, the more space you will need to store it. However, you can build your box in such a way that screws are used to put it together. Back those screws out and store your disassembled box in the rafters when you are done brooding for the year. The only other negative with a wooden box method is that it tends to be heavier than a plastic bin. This is an important consideration if arthritis or lifting heavy items are an issue. People make these boxes in all shapes, sizes, and heights, and they take into consideration the space they have available to them.

Of course, the galvanized metal brooders are a terrific way to go. They have built-in feeders, waterers,

and a heat source. Pretty much worry-free brooding! If you care for them well, these brooders will last decades! Depending on the style and from whom you purchase them, you can simply move chicks down a layer or split them up after they grow to a certain size. If you are wondering how this all works, keep in mind that when chicks are small they can all fit together in the top layer. There are two more layers beneath the top layer, and as the chicks get bigger and the area becomes more crowded, you'll need to provide them with more space. Thus, you'll need to move the chicks to the next layer down. That leaves some chicks in the top layer and some on the middle layer, with the bottom layer empty.

Giving your chicks enough space to grow is the key element here. If you overcrowd chicks, then they can have stunted growth. This is definitely not ideal for meat chickens as they have been genetically selected to grow rapidly.

> As chicks move about the brooder, going from heat to food to water and back, they kick up a lof of dust.

HEAT IT UP!

Chicks in a brooder require supplemental heat. Normally this is what mother hen provides—in addition to her protection and direction involving food and nonfood items. You are going to need to provide this in your brooder. This is honestly going to be the most expensive part of brooding if you choose the cheapest option.

And that option is to use a brood lamp. This often consists of a large metal shield with a socket in the middle along with either a chain or clamp for securing the brooding lamp. The lamps are powered by electricity, and most often people choose high-wattage bulbs that really put out the heat. It is quite common to find 125-watt bulbs used for brooding. Simultaneously, these are a real drag on your power

bill. So that cheap upfront purchase is going to cost you lots and lots in the long haul. And one bulb is not sufficient; if a bulb burns out or breaks then your chicks will be without heat. If that happens and you are not around to act quickly on their behalf, then you could lose chicks.

There are brooder lamps with multiple bulbs out there for purchase. They ensure that if one bulb should go out, then the others still have the chicks covered. However, your power bill will go up accordingly.

So is this really zero waste? Well, let's think about it. Energy that heats the air is highly inefficient. That is what the brooder bulbs do. They heat everything. What is the alternative?

The goal in brooding chicks is to heat the chicks, not everything else. This is where infrared heaters come in to play. There are brooder products that are designed just to heat the animals. The Eco-Glow line, as well as the Sweeter Heater line, are both top products, each with its own design. They are initially a bit more expensive, but the fact that there is a huge savings in the power bill is a big influencer. They pay for themselves rather quickly. The Sweeter Heater can even be put to work in the coop in the middle of winter by hanging it over the roosting poles and turning it on only when the temperature drops to a certain point. Although perfectly designed for the brooder, the Eco-Glow should not be used in the coop as it loses effectiveness below 50°F (10°C). They are fairly easy to clean and disinfect, making it easy to store them when you do not need them. These can be ordered online and are available internationally, as well.

CREATIVE REPURPOSING FOR BROODERS

The brooders mentioned are quick and easy, but are by no means the be-all, end-all of brooders. There are other things out there that people have used for brooders that work equally well. If you have an extra bathroom, then why not brood in the bathtub? As long as you keep shavings and feed out of the drain then you will have no need to call the plumber after your brooding season is finished!

Keeping this in mind, if you have an old bathtub, whether it's metal or plastic or some of the many variations that exist in the world of tubs, you will find convenience in brooding as well as clean-up. You can brood in the garage or perhaps in a part of a shed or barn that is convenient. The hardest part of using an old tub is that covering it can be challenging. Using a very large window screen is still the best method. Additionally, you will need to relocate

your chicks to another container while you perform your weekly clean-out and apply fresh bedding. There is typically plenty of room in an average bathtub for ten chicks to be comfortable for about 4 to 6 weeks, longer if you have Bantam-sized chickens. You will need to cover the drain hole with a piece of wood or plastic to keep chicks from exploring any exit routes as they get older.

Plastic bathtubs are lighter than other tubs and can be raised higher to eliminate the need to bend over more than is necessary. That means much of your brooding equipment and supplies can be stored conveniently underneath your brooder. Space-saving is also a time-saving tip! If you use a plastic tub, you may want to use cinder blocks to raise it up. Some folks build a frame for it out of wood.

REFRIGERATORS AS BROODERS

Got an old refrigerator or freezer that no longer works? Then take off the door and flip it on its back. Now you have a good brooder! You can often find discarded refrigerators or other such items at the dump. Why not ask folks if you can give it a second life?

Again, the surfaces are easily cleanable when it comes to your weekly cleanout. Remove the drawers and shelves and you will find a great place for brooding. Just be sure to give it a really good cleaning because old refrigerators and freezers that have not been used for awhile have the potential to house some ferocious bacteria. Scrub and clean with soap and hot water until you think it is clean enough to eat off of because that is essentially what you are asking your chicks to do. If you think a disinfectant is necessary, then a bleach solution is great to apply for a contact time of 10 minutes before you rinse it off. Remember, bleach is only effective on already clean surfaces.

No matter what materials you choose to use for a brooder, the most time-consuming and energy-consuming task for you will be to clean it out. A brooder that is small and lightweight may be easily lifted so that the bedding can be dumped into the compost bin. If you have a brooder that is big, you may not have to clean it out as frequently, but its weight will cause you to have to scoop out the

bedding. As a timesaver, I have used material to cover the bottom of the brooder that can be lifted and dumped. For example, I have cut poster board to fit down in a bin by cutting the corners diagonally and folding it to fit. When it is time to clean, the poster board can be lifted and dumped to save time and energy. As long as there are sufficient wood chips covering this heavy paper to protect it and it's not damaged by water, it can be reused. However, if it has been soiled beyond use in the brooder, you can save the sheets and use them as a weed barrier at the bottom of a raised garden bed. If your brooder is so big that poster board would not be effective, you may choose other options like a thin rubber mat. Although bigger means you may need two people to dump it, so keep that in mind. If you have the skill set, you may choose to install a hard-wire floor to your brooder where the chicken can walk, but the droppings will fall through. Then you could place a metal sheet underneath to catch the droppings to make it easier to dump into the compost pile. If you're handy, you may choose to build a wire-bottomed brooder right over your compost pile, so you don't have to do anything with the droppings! It will be important to make sure you keep your pile mixed with the right amount of green and brown material as discussed in chapter 3 so you don't torture you chicks with a constant and unhealthy odor.

Coops THAT Will Last

You have a lot of decisions to make when deciding upon a coop design. Will it be pretty or functional, painted or plain, mobile or stationary? How much space do you have to devote to this venture? What about biosecurity?

One of the first mistakes people make is to create a coop interior that is too small. Each chicken will need at least 2 square feet (0.2 square meters) of floor space. Subtract any space taken up by equipment as that does not allow the chickens to use that space on the floor. Hang equipment if you can, or mount it on the wall, to get it up out of the way.

Make the coop tall enough for you to stand up inside it. If it is small, cramped, or requires you to fold up like an accordion to get inside, then you have a bad design. You will not clean as effectively if you cannot get into the space and access all areas. The same principles mentioned for cleaning the brooder also apply to cleaning the coop. Metal, plastic, and painted wood are the best options for keeping things clean.

Wood is a cheap material but red mites love to hide in wood's nooks and crannies. Most coops are made of wood, but people don't take the time to paint the wood on the inside, just the outside. Paint your wood and seal any joints or corners to prevent red mites from finding harborage. Red mites feed on your birds at night and then hide in and under things in the daytime. Most flocks have red mites at some point in their lives. Most people are oblivious, which is why we make it a point to bring it up as a measure of good management. Periodically, you should take your coop apart and dismount the equipment from the walls to do a full clean-out. This should be done once a year.

If you plan on putting insulation into your coop, then be sure to cover it with wood so that the chickens will not tear it up or eat it. Chickens love to explore new things. Insulation of any sort is just one more thing to be explored without any regard to how much money you may have spent on it. Keep your insulation intact in order to have it do its job for you.

Your coop can be made of different repurposed materials; however, make sure that they can bear the weight of any materials you should hang for feed or watering purposes. Additionally, mounted nest boxes should not strain against the walls or go all the way through the wall causing gaps where drafts can come in. What if you're not sure if the coop you have built is tight? Then use a nontoxic smoke emitter just like they use in commercial poultry houses when they want to see if their ventilation system is working properly. You can buy five for about $20 and they will burn for almost 5 minutes. That will give you enough time to close up your coop and walk all the way around looking for where cold air can get in and harass your chickens! Of course, do not do this with your chickens inside, or you either, for that matter! Just Google these nontoxic smoke emitters and buy them as needed. Inspect USA, a company that sells smoke emitters, among other things, carries just what you need.

MOVING AND GROOVING VERSUS STANDING STILL

Is a stationary or mobile coop design considered more low waste? That depends on you. Not all coops are meant to stay put, so if your intent is to supplement your chicken feed with fresh grasses from your yard while spreading fertilizer throughout your yard, a mobile chicken coop may be the best option. If moving a coop from one location to the next on a daily basis causes stress and strain, then you may prefer a stationary coop. If possible, set up your coop so it has four separate pastures for you to rotate through. If you can manage more, then go for it! Most coops are stationary and have only one run, which quickly becomes denuded. That is why several runs or pastures are recommended, so that the ground does not become denuded. Mobile coops can simply be pushed or pulled to fresh grass or pasture. Their drawback is often their weight and the fact that the ground predators can sometimes dig under them and eat the chickens, a sad way to lose your resources. Another drawback is that tires can go flat unless they are solid rubber.

Chickens quickly wear out the ground around their coop. To prevent a build-up of organic material that can turn into run-off into local waterways, you will need to learn how to manage the land and not just your chickens. Nobody who is raising chickens in their backyard wants to add to the burden on their local environments. So putting up gutters to redirect water to move it away from soiled outdoor spaces is a good idea.

Planting different materials in different pastures is another method to try to draw out the nutrients deposited in chicken manure. For instance, in pasture 1, you can have grass mixed with white clover. In pasture 2, you can have another type of grass mixed with clover. In pasture 3, you can plant something taller, like sunflowers or sorghum or corn. In pasture 4, you can plant alfalfa or vetch. This way, the chickens get to explore a different pasture each day and, if the land is well managed, experience different forages.

The more space you can make for your chickens while still maintaining biosecurity, the better. To rest a pasture for nearly a whole season while you let sunflowers or corn grow is not out of the realm of possibility, but it will make you more aware of the pressures that animals put on the soil that we share with them.

So if you are building a coop from scratch, then consider multiple outdoor runs. You can also build a mobile coop. This does not, however, remove the learning curve with regard to pasture forage management. You see, chickens are not ruminants, so that means that most older grasses are not available for them to fully utilize like a cow or goat does. Rather, legumes should be your goal. Clover or alfalfa are wonderful forages and once they are trimmed back by mowing or chicken consumption, they often regrow or even bloom. You can also harvest alfalfa every 45 days or so, and once it's dried, can store the hay for use in the winter months.

Options FOR Fencing

Now let's talk about fencing. If you are operating on a smaller budget, then you can purchase solar-powered electric fencing. This fencing is turned off, rolled up, and deployed again around the space to which you move the mobile coop. This takes time and effort. The fencing is awkward to lift when it's all rolled up, but all in all, it is a lightweight alternative. On the downside, it will not keep out avian predators.

If you are using a movable coop, you can either move the fencing each time you move the coop or not. Your other option is to purchase enough fencing to move a coop around within the fenced space without having to relocate the fencing each time. The coop moves once a week within the pasture to one of each of the four quadrants. Then once the month is up, the fencing is moved to the neighboring section of ground and the process starts over again.

Any mowing to the ground that needs to be done on land that was recently vacated by your chickens is done shortly after moving them. If you have additional plots of ground awaiting your coop and chickens, that may mean you have to mow the other plots before your chickens ever set foot on them. Chickens tend not to eat tall, rank grass. Consuming such grasses may cause the chickens to develop an impacted crop. Now there is a vet bill that absolutely nobody wants and is completely preventable.

The next question that everybody asks is what to plant. Well, there are no good answers yet. First, you want to plant something like clover that the chicken can actually digest and utilize. Beyond that, there is very little information about the perfect forage blend to use. Why?

There is no pastured poultry industry of sufficient size, as of yet, to do the rigorous and systematic scientific research that is needed to answer questions on productivity. In addition, plants that grow well in one region may not do well in another. So what works for one person may be inefficient for another. Change the breed or environmental conditions and you have to start your experiment all over again with the new parameters. It will take years to get the answers right for the majority of farmers so that they do not lose their shirts in the process, and they deserve answers that will benefit them. But right now, they do not have a strong enough voice to be heard nor do they collectively make enough money to pay for the work to be done. Besides, a balanced diet is what a chicken needs to grow best, and pasture is viewed as a nice supplement rather than a full dietary replacement for growing chickens.

FUNCTIONALITY CONSIDERATIONS

Mobile coops can get heavy very quickly. Even adding a water bucket can make a mobile coop so heavy for some folks that they can no longer move it easily. Tire weight capacity is another issue. Pneumatic tires can go flat if punctured of overburdened, which negates the whole point of a mobile coop! And it never seems to happen at opportune time either. Murphy's Law says you will get a flat tire on your coop when you are in a hurry and trying to move the coop before a giant rain cloud dumps all around you!

There is such a thing as solid tires. They are all rubber and more expensive but well worth the investment if you can afford them. The drawback is that they do add a little more mass to the overall weight of your mobile coop. This is an issue if you are moving coops with good old-fashioned human power. The alternative is to use a tractor or perhaps even a riding mower to move a coop.

Zero waste can also mean capturing that which causes a mess. Let's talk about rainwater for a moment. There are all kinds of catchment systems out there for homes and other structures, but few are available for coops. If a catchment system can keep water clean and also return it to the chickens for their consumption, then that is a win-win situation. Some systems dispose of the first flush of rainwater, which often contains dust and wild bird feces from the coop roof. The system's pipe type to redirect the water into catchment containers should not allow mosquitoes to lay eggs and breed within grooves. Mosquitoes can be carriers of fowl pox, a disease affecting birds, including chickens.

So if a catchment system existed that removed the first, flushed material and kept out mosquitoes, then its next challenge would be to make the water accessible to the chickens themselves. A float for a nipple drinker–style waterer would keep that waterer system as closed to dirt and other airborne contaminants as possible. Closed water systems, like nipple drinkers, also mean the chickens cannot defecate in the waterer and foul the contents.

THE Waterers OF Yesterday

The old Bell- or Plasson-style waterers could also work; however, they could be bumped by the birds so that they spill, wetting the litter and bedding. That led to higher ammonia levels. Also, they had to be dumped and scrubbed daily, which is an additional step to chicken keeping. They had to be disinfected weekly, which was a procedure unto itself. Since this type of waterer could be defecated in and collected dust from the air, it was an open-water system that brought an increased risk of disease with it. By the 1990s, the commercial poultry system moved to a closed-water system to improve the overall health of flocks and reduce disease pressure on the birds.

Not all roofs lend themselves to rainwater catchment, but some do. Some flock owners may just want to prevent the water from gathering in the pen and making a mud bowl. Other flock owners may wish water to enter the gutters and then flow harmlessly away from the coop into a field or to water thirsty plants in a yard. Striking up a conversation with a contact at a local gutter company may yield an opportunity to purchase low-cost fragments of gutter left over from a job where product quantity was overestimated.

Another functional consideration is weight in a mobile coop design. Everyone will need to think carefully about those for whom the responsibility for daily coop movement lies. Can this person lift and pull or push a mobile coop if a full tube feeder carries an additional 35 pounds (16 kg) of weight? It is certainly more convenient to move the coop when the water storage container is almost empty. Every gallon of water weighs about 8 pounds (3.5 kg). If you have a 5-gallon (19 liter) bucket as a reservoir for water plus a large tube feeder, then that is about 75 pounds (34 kg) of weight. When the grass is growing perfectly for the chickens (it is young, tender, and has new growth of less than 2½ inches [6 cm]), it may not be convenient to wait for the water to run low and feeder to empty out just so you can finally pick up and move the coop once more. Take the time to examine the different equipment options from multiple outlets in order to both spend your money wisely and also to meet the needs of your personal coop.

DAY RUNS WORTH CONSIDERING

Day runs are a popular option for moving chickens into a certain area of the farm or yard. This is often an open-topped structure that does not prevent the chickens from flying out or avian predators from flying in. It does keep the chickens corralled, often right where you want them to apply their manure and scratching capabilities. The sides are often flimsy and could be upended quickly by a ground predator such as a coyote or fox. Sometimes these day runs are covered, but usually only by a wire top rather than a solid top that would prevent wild bird feces from coming into contact with your birds. Day runs are best employed under supervision while you are working in the yard with the chickens. That means if a predator should enter the scene, it can be run off by the flock owner or the chickens can quickly be put back into the safety of the coop.

EXAMPLES OF REPURPOSED COOPS

There are no perfect versions of a coop out there. Make it easy to clean or move around in and most folks are happy. Dog houses are great for two or three chickens. Some are wood, which can be painted, or are plastic, which makes them easy to clean. Flock owners who live in areas that experience very cold winters may want to consider if insulation can be added, but it may not be needed in areas where it does not get as cold in the winter. A repurposed dog house will still need to keep birds cool in the summer. That is where shade from corrugated plastic roof panels can help. If the dog house is in a dog run, then putting a cover over the top is relatively easy. It does not make the coop mobile, but it does provide protection from avian predators and wild bird feces. Digging predators will still need to be abated so adding chicken wire flush with the ground at a 90-degree angle will be a deterrent to most predators.

If you are especially handy, then you may be able to look at one man's junk and see a whole new possibility for its second life. Anything with wheels and a sturdy, functional chassis can fall in that category. Take a trailer, wagon, RV, or bus and you can clean it out to prepare it for a coop. The biggest challenge will be removing surfaces that cannot be cleaned and replacing the surfaces with those that can be. Secondarily, you will need to remove anything sharp that could injure or puncture the chickens. Holes will need to be closed up, as well.

If your chassis has a motor in it still, then you will need to consider if you want to have it be a mobile or stationary coop. Will it move on its own? If not, then consider taking out the motor and removing any oil or gas that can leak into your soils.

An old trailer with two wheels is perfect to repurpose once the interior is cleaned out. Pop holes or doors for the chickens to use will need to be cut into the side and a ramp to the ground will be needed. The doors will need hinges and locks for keeping out the weather and predators. Also, door size is important. Some dominant chickens guard doorways and let in only certain members of their flock. Having several doorways helps to eliminate this dominancy problem and has the potential to add some cross ventilation. Additionally, old trailers are usually tall enough for a person of average height to stand up in, making them all the more convenient. You will, however, need something to pull them with around the farm as well as a tire gauge for monitoring tire pressure. Flat tires are an absolute drag!

OTHER THINGS TO THINK ABOUT

Nest Boxes

A rule of thumb with nest boxes is that they should be large enough for a single hen to fit in to be able to lay her egg every day. Bedding or a plastic AstroTurf mat in the form of a nest box liner should be inside to help keep the eggs clean. Do not let hens sleep in nest boxes or they will defecate in them. So a perch along the front of nest boxes is great if you can lift it up and block the entrance at night. Nest boxes should not be mounted to the wall higher than 2 feet (61 cm). Why? Hens may have hard landing when they jump down from the nest boxes and that can lead to injury. Keep your bedding in the coop several inches thick, and you can help prevent such injuries.

Red mites love to hang out in nest boxes and bedding. So nest boxes made of plastic, painted wood, or metal are best for cleaning. Build or purchase your nest boxes of these materials to ensure that you keep red mites at bay. Regular quarterly disassembly and cleaning of nest boxes is a must along with weekly replacement of bedding in the boxes.

Did you know that galvanized metal nest boxes can be sold by the number of holes? Yes, indeed! That means you need to buy only as many nest boxes as your flock needs. Every hen does *not* need her own nest box. Rather, you will need one nest box for every four hens. You heard that right! And they will be quite happy to share. Watch carefully for broody hens taking over one of those nest boxes because that can cause hens to lay on the floor if they cannot gain access to a nest box. A colony-style nest box is a nice alternative. It is the same depth and height as that of a regular nest box, but the length is much longer. It will allow several hens to enter and lay at the same time.

Rollaway is the way to go! Slant the floor toward either the front or back to allow an egg to roll out of the nest box and away from nosy beaks. Using plastic nest box pads makes cleaning a rollaway nest box much easier. Open the flap and you will be pleased to see all the eggs gathered together, clean and ready for pick-up.

The darker the better! Hens prefer nest boxes that are dark. So mount your boxes in the darkest part of the coop. Or, if that is not an option, orient your coop so that the nest boxes are not necessarily in direct sunlight from a window. Sometimes, to prevent floor eggs, you may need to hang curtains across the front of nest boxes to make them more appealing to new hens. Vinyl curtains are easy to power wash during the cleaning process.

There are many cute, simple, low- or no-cost methods of making your own nest boxes. However, they usually end up being more work in the long run. From milk crates stacked atop one another, to plastic buckets with perches, to homemade wooden

Perches to Consider

Perches or roosting poles inside the chicken coop are for roosting at night. Chickens like to be up off the ground and near the roofline. That is unless it is so unbearably hot in the coop that they cannot roost up there. Each chicken tends to radiate as much heat as a 60-watt lightbulb, and we all know that heat rises. That means that the perching area can be the coziest part of the coop in winter! If possible and if necessary, you can mount your infrared heater in the coop over the perches to keep your chickens nice and warm in the winter months.

Perches can be made from metal, plastic, and wood, which is the most common material used. There are pluses and minuses to each, but folks are going to most easily find and use wood for their backyard flock. You want perches that can handle the weight of several 6.5-pound (3 kg) chickens on them for many hours. The bigger question is the shape. A small perch makes a chicken wrap its toes all the way around the roosting pole. In winter, this could mean that a chicken may not be able to lift up its breast feathers and cover its toes to keep them warm. Frostbite of toes trying to hang on to the pole and getting too cold can result. Use a pole that is rounded rather than square or T-shaped. Wood is warmer than plastic or metal. If you can get a pole that is 2 inches (5 cm) in diameter, then that is a terrific size for your chickens.

nest boxes, you should definitely consider mounting them to the wall at an angle to allow them to roll the eggs away from the entrance. If an egg rolls out and breaks on the coop floor, not only is it a mess to clean up, but a chicken that figures out it is tasty can start egg-eating behavior. Egg-eating behavior is not curable. It can cause a great deal of loss in product if a chicken insists on breaking her own eggs as well as the eggs of other hens to get to the yummy contents. Possibly the chicken itself will need to be removed from the flock to prevent it from teaching others its bad habit.

Perches need to be at minimum 18 inches (46 cm) off the ground. Then stagger the perches back by 6 inches (15 cm) and up by 12 inches (30 cm) so that birds that are more dominant and roost higher up will not defecate on the lower birds at night. If you space the perches too far apart, then birds may not be able to hop down safely and accurately, which can lead to a fall and injury.

Painting perches may seem silly, but during clean-out, it is a great time-saving step. On a weekly basis, run a gloved hand over the entire surface of the perch in order to identify any areas that could poke your birds when they walk along the perch. An injury to the foot can lead to bumblefoot and perhaps to permanent damage due to scar tissue.

There are many examples of coops with scavenged wood from multiple sources. Recycled barn wood, pallets, leftovers from building projects, auction finds, Craigslist, and fencing companies that have removed old fences before putting up new ones all can be sources. They can be pieces from a fencing project or perhaps wood from a fallen tree. It is not recommended that you use treated wood for perches. Check your chickens monthly for signs of a crooked breastkeel bone if you are repurposing branches from your property as perches.

CHOOSING A FEEDER

Feeders are designed to make food accessible to chickens at all hours of the day. That means that the height of your feeder, which is set at the height of the chickens' backs, will need to be adjusted upwards as your chickens grow. There are many great designs that either hang or mount on the wall in order to free up floor space inside the coop.

Of course, if you choose to hang or mount a feeder, you will need to make sure that the hardware can handle the weight of the feeder as well as that of the feed. Use a chain with an S-hook to adjust the height of a hanging feeder. Feeders mounted to the wall should have a cap over the top to prevent chickens from defecating inside should they find it a convenient, or even accidental, perch.

Feeders that sit on the ground should still be placed on blocks that can be used to lift the feeder to the height of the chickens' back. It is a good idea to place a roosting guard over the top. Also, if you're putting a feeder on the ground, you will need to design it in such a way that the chickens cannot scratch the feed out with their feet. Feed that falls on the floor gets mixed into the bedding and is wasted.

By putting the feed at the height of the birds' backs, you are reducing a problem called billing. Billing is when a chicken, or another poultry species, scoops feed out of the feeder with its beak. This is a side effect of boredom. Chickens that are billing are simply trying to fulfill their biological need to explore their surroundings for food and other interesting things. Sixty percent of a chicken's day is spent exploring her surroundings!

Feeders must be cleaned weekly. The most common types of feeders are plastic and metal. Those are easy to clean and dry. Galvanized metal tube feeders can hold up to 35 pounds (16 kg) of feed in some cases. PVC pipe or feeders in plastic buckets are also easy to clean. You can even spray them with disinfectant weekly before putting them back to work in the coop.

WATERERS

The principles for waterers are pretty much the same as that for chicken feeders. If you are using an open watering system, then you will need to clean and scrub the base daily. You will also need to set the waterer at the height of the chickens' backs.

For the overall health of the flock, it is recommended that you use a closed watering system. This is the nipple drinker that is so ubiquitous in backyard chicken keeping. Not only does the water stay cleaner, but it will save you many hours of cleaning over the course of a year. Nipple drinkers can either be set into the side or bottom of a container. Side nipples are rarer, and if they're mounted 1 to 2 inches (2.5 to 5 cm) above the bottom of a container, will not clog up with debris carried in with the water.

The amount of bacteria in a closed water system takes a month to reach that of what an open watering system takes to accumulate in a week!

Nipple drinkers must be set at a different height, however, to keep the chickens from wasting the water and wetting the litter. Wet litter leads to ammonia production, which nobody wants! A chicken should be able to reach up and activate the nipple tripper and let the water flow down into its beak. So the nipple drinker should be set higher than the average height of the chickens' heads.

Check nipple drinkers weekly by triggering the nipple. You can then see if it is working properly. Also, be sure you can get into the waterer and scrub the inside monthly. Eventually the nipples will need to be replaced. The seals can go bad and start leaking. Every day you should look under the waterer to see if you notice a slow leak.

DUST BATHS

Dust bathing comes naturally for chickens. If you do not provide a space in which to dust bathe, then you may begin to find that parts of your chicken run will begin to develop craters. These craters are simply spots that the chickens have chosen to use for dust bathing. Chicken will kick up loose dirt and work it through their feathers to remove old oils on their feathers. You can provide your own dust bath in the coop by bringing a container filled with loose dirt or potting soil.

If you have hardpan soil or if the run is on concrete, then your chickens will dust bathe in the shavings and litter inside the coop. This will kick up a tremendous amount of dust. Allow your chickens to dust bathe two times a week and they will be quite pleased with the change of pace. Do not let your dust bath material get wet, or you may find that mold or bacteria will grow. You wouldn't know this unless you take a sample to a lab and pay to have it done, but no one's going to do that.

Something as simple as a plastic bin will work. Keep in mind that chickens will kick some of the material out of the dust bath and into the litter. A plastic masonry mixing tray or kitty litter pan will do the trick for a small number of chickens. If it gets much larger, then lifting the container in and out of the coop becomes difficult. If your run is sufficiently large, then you may consider permanently installing a container with a lid that you leave open twice a week to allow your flock to have access.

There are no perfect coops or coop components. You do, however, have lots of options for customizing your coop. Now you have the basic principles you need in order to have greater success in cleaning your low-cost or no-cost coop.

So go ahead! Pull apart those pallets you saw on the side of the road and put them to work on the inside of the coop. But be sure to check whether they have been treated (see page 62 for more). And give them a nice coat of paint so that you find cleaning to be a snap! There is no need to worry about mites so long as you make absolutely sure you seal the cracks and paint the wood!

Lining Your Nest Box

Always choose plastic nest-box liners because they are reusable. You should have enough on hand to fill each nest box plus extras. You can pop one of the extras into place when one of the original liners is soiled either with droppings or a broken egg. All you need to do is scrub it clean in a bucket of soapy water, rinse it off, and then let it sit in a bucket of disinfectant. Let it sit until the contact time noted on the disinfectant bottle is up, and then rinse off the disinfectant and let it dry fully.

WHEN **Hens Stop Laying**

Your farm has a natural ebb and flow. There are seasons, of course, but we're talking about the longer-term ebb and flow. You have new projects, farm maintenance, and the periodic assessment of whether to scrap or abandon a project or try another tack.

The day will come when your chickens are no longer laying at a rate that makes sense for your farm. Whether it is for profit or for your breakfast table, you will get to the point where you need to think about starting anew. Not everyone has the money, time, or desire to deal with the biosecurity headaches of keeping two separate coops.

It may be that you need to remove the old flock and replace it with a new set of hens. You could sell them as spent hens at a local auction and maybe get a couple of dollars per bird. You could sell them to a family that just wants chickens, but are not interested in the eggs, but that kind of a family would be hard to find. So that means you will probably be processing the chickens.

You could do this yourself and either sell them or put them in your own freezer. You could take them to someone in your state, which will likely cost somewhere between $5 to $10 per bird. You could

turn right around and sell them live as spent hens and have buyers process them on their own. Regardless, when it comes to selling eggs and chicken meat, you *must* check with your state's laws to be sure you are not violating the rules. Start with asking your local Cooperative Extension office. They may refer you to someone at your state's land grant university. If you reach a dead end there, contact someone who works with eggs or meat at your state's department of agriculture.

These will not be the young and tender birds that you buy at a grocery store, and they will taste different. An older hen has had time to take on the flavors of age and that will be reflected in the taste of your dishes. Older hens are tough! You cannot just plop them in the oven and cook them like you would for a broiler or roaster. You need to cook them low and slow over a long period. That is why they call them stewing hens!

"Meating"
A NEW Flock

Some people like the idea of raising their own chickens for meat production. They like feeling that they can be done with a meat chicken project in just a few weeks. Honestly, if you are raising a fast-growing strain of Cornish cross, then you can be done in as few as 6 weeks.

A Cornish game hen is still a Cornish cross but it is processed when its live weight is only about 2 to 3 pounds (1 to 1.5 kg). That happens when it is about three to 4 weeks old. Keep going until a chicken is about 6 weeks old and you will have a nice broiler on your hands. Keep going another 2 weeks and if your chickens get to be over 6 pounds (2.7 kg), then you have a nice roaster.

Going much longer than 8 weeks and you will need to be an excellent manager to keep your chickens alive to a higher weight. But that is entirely possible if you are raising a slow-growing strain of chicken.

They take about 8 to 10 weeks to reach the size of a broiler and even longer, maybe even 15 weeks to reach the size of a roaster. That also means you are going to be paying the feed bill for longer. These chickens will be less efficient at converting feed into muscle. So you will need to watch the weather, keep the pastures watered if pasturing your slow growers, and plan to weigh a few of them every 3 weeks to see if you are still on track.

> Roosers can be raised for meat and hens raised for eggs.

PROCESSING EQUIPMENT

If you are the intrepid chicken keeper who likes to do everything yourself, then there is no doubt that you are going to be processing chickens at some point. You are going to need some key pieces of equipment to make this all work well. First, you will need a sharp boning knife. You can actually make all this work with just a sharp boning knife and a pot of scalding hot water, but we will go further into pieces of equipment that make the process easier and less of a chore.

So let's start with the killing cone. You can make one by fashioning a funnel out of a piece of flexible metal and hanging it on a nearby tree or other building. The no-waste method would be to take an abandoned (not stolen, note the difference) road cone and hang it from a tree pointed downward. You may have to widen the point by taking a pair of tin snips to the narrow end. This will allow a chicken's head to come out the bottom without overstretching the bird.

The boning knife can be used for cutting the veins and arteries of the neck. A clean bucket can be used to collect the blood for blood sausage. A pithing knife can be used to pith the bird before you move to the scalding step. A pot of scalding hot water that is at 140°F to 150°F (60°C to 66°C) will be used to scald the feathers in preparation for plucking. The bird needs to be in the scalder for only 30 to 60 seconds, and then it will be ready for picking. Check by pinching the scales on the leg to see if they come off easily. Pick the bird starting with the largest feathers of the wings and tail before moving to the breast and the rest of the body. Or you can buy a picker or chicken plucker and be done in 15 seconds! This is the route we like!

Then you'll move the bird to a table where you will remove the head, legs, and gastrointestinal tract. After you are done, you will need to keep your chickens in an ice bath for up to four hours so that the thickest part of the breast muscle reaches 40°F (4°C) or colder. Then package your chickens in large bags and either put them in the freezer or get ready for cooking.

Keep the legs after removing the scales because it can be used for soup stock. The giblets are in the gastrointestinal tract and can also be kept. The giblets include the liver, gizzard, and heart. Sometimes people lump the neck in with the giblets, but those are best frozen separately because they have many uses.

What do you do with the remainder of the gastrointestinal tract, head, lungs, and preen gland? You can compost them, of course! Composting is a great way to recycle these key nutrients back into your soil. Just be sure to bury them deep enough that scavengers will not dig them up and steal from your compost pile.

Taking Stock

Making delicious chicken stock is very much a lost art. For this project, you will be doing two things at once. If you have held on to your chicken necks and feet, then get them out of the freezer. When you defrost one of your hens, you may choose to cut it up. If you debone any parts of the bird, then toss those bones in with the parts for soup stock. You can crack the bones of the legs open to help the stock gel. I also like to slice the skin of the legs at intervals along the toes and leg to allow collagen to be released in the cooking. Toss in some vegetable parts that you may have been saving, as this is a perfect use for these. That base of broccoli or maybe some onions or carrots can add additional flavor.

Cover the parts with enough water so that they are submerged in at least 3 to 4 inches (8 to 10 cm) of water and turn the heat on your stovetop to medium-high. Once the water boils, turn it down to low so that the stock simmers only slightly; leave it to simmer for about 4 hours or longer. Skim any fat that comes to the top. About halfway through, I like to add salt and pepper to taste, but earlier in the process you may choose to add garlic or a bay leaf. You can then pick any meat off the neck bones and discard the other bones.

Once your stock is made, you can save it in freezer bags for later use. Need smaller amounts? Freeze some stock in ice cube trays and, once it's frozen, remove them from the ice cube trays and store in freezer bags in the freezer. Then you only need to grab one or two cubes as needed to add to recipes later. If you are feeding a crowd now, chop some fresh veggies, add some chopped chicken or noodles, and let simmer for about 30 minutes. Bon appetit!

LURE THEM IN

Chicken feet are not for everyone. Let's face it, in the United States and Canada, the vast majority of people do not know how to cook with chicken feet and legs. But that does not mean you should throw this away during processing! That would be a huge Waste Alert! Save the legs and sell them or use them. How can you use them, you ask?

Why, as bait, of course! Chicken legs and necks are perfect crabbing bait! Got a small creek or a lake nearby? What about crawdads or crayfish? They like chicken as much as you or I do! So start planning how to make a crab pot or crawdad trap and make this a fun project to do with your kids!

Now aren't you glad you kept those scraps of chicken wire that you thought had no good use? Just avoid that crab pinch and put them in a pot! You can even stretch your chicken efforts further by making stock from the crab or crawdad shells! Now that really is maximizing your efforts!

Resources and Acknowledgments

Resources:

Written Materials

Chicken Whisperer Magazine

The Chicken Whisperer's Guide to Keeping Chickens

Products & Equipment

The Sweeter Heater, www.sweeterheater.com

The EcoGlow Chick Brooder, www.brinsea.com

Pasture in a Bag, www.chaffhaye.com

Organizations

National Poultry Improvement Plan (NPIP), www.poultryimprovement.org

National Research Council (NRC), www.nationalacademies.org/nrc

The United States Department of Agriculture (USDA), www.usda.gov

National Organic Program Standards, www.ams. usda.gov/about-ams/programs-offices/national-organic-program

Acknowledgments:

From Andy Schneider

After hours of looking at a blank piece of paper, I have come to the conclusion that there are truly no words that can express my gratitude to my beautiful wife, Jennifer. Thank you for all that you did on this book and all that you do in my life. I love you.

I would also like to thank my coauthor and friend, poultry scientist Brigid McCrea, Ph.D. We have worked together on various projects for over 9 years now, and it has been a complete joy working with her. Her passion for poultry education is second to none, and I am honored to have worked with her on this project. She has helped mold the Chicken Whisperer® brand into what it is today: a place where people can find science-based, fact-based, and study-based information to help them raise the healthiest flock possible. Thank you, Dr. McCrea, I could not have done it without you.

About the Authors

Andy Schneider, better known as the Chicken Whisperer®, has become the go-to guy across the United States for anything chickens. He is the National Spokesperson for the USDA-APHIS Avian Health Program; Editor in Chief of *Chicken Whisperer Magazine*; author of *The Chicken Whisperer's Guide to Keeping Chickens* and *Chicken Fact or Chicken Poop*; and host of the very popular *Backyard Poultry with the Chicken Whisperer* radio show and podcast.
He has been featured on CNN, FOX, ABC, NBC, and CBS, as well as in *Time* magazine, *The Economist*, *The Wall Street Journal*, *Atlanta Journal-Constitution*, and countless other publications. Andy has spent the last decade traveling the country with his family, spreading the chicken love by providing poultry workshops, speaking engagements, book signings, and on-sight educational visits. Last year, Andy and his wife, Jennifer; son, Caleb; and daughter, Lily, purchased a 13-acre homestead just north of Atlanta, where they raise chickens, turkeys, cows, goats, pigs, and rabbits. They have also started the Chicken Whisperer Farm School, where they teach children of all ages about agriculture.

Brigid A. McCrea, Ph.D., began her lifelong love of chickens quite by accident, and she attributes it all to her involvement in 4-H. With the help of the family mechanic, who was also a show-chicken breeder and a 4-H poultry leader, she began to raise and show chickens. The poultry community is full of kind people willing to share their knowledge with fellow poultry enthusiasts, and it was this camaraderie that spurred her on to study poultry and birds in college. She received her B.S. and M.S. in Avian Sciences from the University of California, Davis, and then received her Ph.D in Poultry Science from Auburn University. Brigid is pleased to share her poultry knowledge with everyone who is interested in learning how best to start or serve their flock.

Index